Physical Principles Explained

D0233591

Titles in the series
Electrotherapy Explained
Physical Principles Explained

Physical Principles Explained

John Low BA(Hons), MCSP, DipTP, SRP
Former Acting Principal, School of Physiotherapy, Guy's Hospital, London

Ann Reed BA, MCSP, DipTP, SRP
Senior Lecturer, Institute of Health and Rehabilitation, University of East London

Butterworth-Heinemann Ltd
Linacre House, Jordan Hill, Oxford OX2 8DP

ℛ A member of the Reed Elsevier group

OXFORD   LONDON   BOSTON
MUNICH   NEW DELHI   SINGAPORE   SYDNEY
TOKYO   TORONTO   WELLINGTON

First published 1994

British Library Cataloguing in Publication Data
Low, John
  Physical Principles Explained
  I. Title   II. Reed, Ann
  615.8
  ISBN 0 7506 0748 3

Library of Congress Cataloguing in Publication Data
Low, John (John L.)
  Physical principles explained/John Low, Ann Reed.
  p.   cm.
  Includes bibliographical references and index.
  ISBN 0 7506 0748 3
  1. Physics.   2. Physical therapists.   3. Medical physics.
  I. Reed, Ann, SRP.   II. Title.
  QC23.L866   1994
  530–dc20   93–33995
                                                     CIP

Typeset by TecSet Ltd, Wallington, Surrey
and printed and bound in Great Britain by Clays Ltd, St Ives plc

Contents

Preface

The intention of this book is to provide a framework for understanding the scientific principles which underlie some aspects of physiotherapy practice. It is particularly intended to support some of the basic physics needed to deal intelligently with the therapeutic application of physical agents, as discussed in *Electrotherapy Explained: principles and practice*.

The approach used in this book is essentially non-mathematical. This is not to decry or deny the value of a mathematical basis; it is only an intention to communicate the underlying concepts by another route.

It should be recognized that the use of similes and analogies, which we have exploited in this book, is a limited way of addressing some of the important ideas presented. They should be regarded as rungs on the ladder of understanding. Nevertheless, we hope that we are explaining these important and often difficult ideas in a manner that can be embraced by all. In this connection it may be noted that this book is written without assuming any significant prior knowledge of physics. We have tried to utilize a simple and comprehensible approach while tackling some of the fundamental mechanisms of the physical world. We have not written on the detailed structure and working of therapeutic sources. This book is in no way a users' guide or repair manual. Some accounts of the work of the well-known scientists of historical importance mentioned in this book are to be found in sources in the bibliography.

In an address, Richard Feynman suggested three ways in which science is of value: firstly, that knowledge is an enabling power allowing things to get done or be produced – it can, of course, be a force for good or evil. Secondly, there is the fun, the enjoyment obtained from reading or learning about science – the 'wonderousness' of it all. This is something he notes that is not celebrated as are music, art and literature. The third value he suggests is the discipline of recognizing a degree of uncertainty – a freedom to doubt in all things.

In our limited field we would be happy if in any small way this book contributed to an understanding of the usefulness, the fun of and the uncertainty of science.

We are very sensible of the support we have received from others in the preparation of this book and of *Electrotherapy Explained*. We would like to acknowledge the stimulation of our students, the cooperation of our colleagues and the forbearance of our families in these enterprises. Further, we would especially like to acknowledge the debt we owe to Caroline Makepeace of Butterworth-Heinemann for her efficiency, encouragement and endless patience.

John Low
Ann Reed

1. *Energy and matter*

Force
Systems of measurement
Energy
Conservation of energy
Kinetic and potential energy
Measurement of energy

The physical world may be conveniently described in terms of matter and energy: the former occupies space and has mass while the latter is the capacity of a system to do work. Physics is the study of energy and matter and their relationships. For such study it is necessary to have a means of describing both matter and motion in some agreed units.

There are three fundamental concepts, measured in internationally agreed units, from which many others can be derived:

1. The concept of length or distance, measured in metres.
2. The concept of amount or quantity of matter, measured in kilograms.
3. The concept of time, measured in seconds.

From these fundamental dimensions many other units can be derived. For example:

Length × length = area in square metres, e.g. 2 m × 2 m = 4 m^2

Length × length × length = volume in cubic metres,
e.g. 2 m × 2 m × 2 m = 8 m^3

$$\frac{\text{Mass}}{\text{Volume}} = \text{density, e.g. } \frac{16 \text{ kg}}{8 \text{ m}^3} = 2 \text{ kg m}^{-3}$$

N.B. (The notation 2 kg m^{-3} means 2 kilograms per cubic metre. Similarly, 2 m s^{-1} means 2 metres per second. This form will be used throughout.)

$$\frac{\text{Distance}}{\text{Time}} = \text{speed, measured in metres per second, e.g. m s}^{-1}.$$

$$\frac{\text{Distance in a uniform direction}}{\text{Time}} = \text{velocity, measured in m s}^{-1}$$

$$\frac{\text{Distance}}{\text{Time} \times \text{time}} = \text{acceleration, measured in m s}^{-2}$$

While the first three examples describe matter, the next three are concerned with motion and lead to a most important concept – that of force.

FORCE

Force is what produces or tends to produce motion. In simple terms forces are pushes and pulls. The idea of force is central to much of what is to follow in connection with electrical potential, mechanical pressure and much else.

Newton's first law of motion tells us that a force is needed to alter the velocity of a body, i.e. to change its speed or direction of motion. This means that if a body is stationary the forces acting upon it must be balanced with one another and the system is said to be in equilibrium. If the body is moving it will continue to move at a constant speed until a force causes it to stop, slow down, speed up or change direction.

Moving bodies possess a quantity of motion called *momentum*. It is equal to the mass times the velocity of the body ($m\,v$).

Newton's second law of motion explains that the rate of change of momentum of a given body is proportional to the resultant force acting upon it and takes place in the direction of the force.

This is simply an extension of the first law and means that, for example, a large force acting on a small object will accelerate it rapidly (say, a hefty kick of a light ball) whereas a small force acting on a large object will accelerate it slowly.

Rate of change of momentum can be written as mass (m) times acceleration (a). Expressed in symbols, this is:

$$F = ma$$

Force (F) is measured in *newtons* (N). The force needed to accelerate a body of 1 kg mass at 1 ms^{-2} is 1 N. If a force is applied to a 1 kg object such that at the end of 1 second the object has a velocity of 1 m s^{-1} and at the end of the next second it has a velocity of 2 m s^{-1}, and so on, then that force is 1 N. Similarly, a 1 kg object decelerating at 1 m s^{-2} would exert a force of 1 N.

$$N = kg\ m\ s^{-2}$$

Notice that force, like velocity, is a *vector quantity*, which means it has *direction* as well as magnitude.

SYSTEMS OF MEASUREMENT

For dealing with and communicating about the physical world, systems of measurement are necessary. There are three aspects of measurement: the numerical value, the unit and the level of accuracy. For example, a piece of cloth might be described as 2.5 m long to the nearest cm.

The fundamental units of length, mass and time are entirely arbitrary, being chosen from a variety of human-sized measures which developed from practical use. Some of the names used in the old imperial system – feet and stones – make this very evident. The basic units are those that can be conveniently handled by humans. Thus a kilogram is the sort of mass that is often held in the hand and a metre is approximately the distance from one shoulder to the opposite outstretched hand, a posture often seen in unravelling a bolt of cloth or roughly measuring a length of flex. A second is approximately the interval between the beats of the resting human heart rate.

Actually the human body was widely used as a convenient yardstick during recorded history. The distance from the tip to the interphalangeal joint of the thumb is well-known as the basis of the inch, but palms, fingers and cubits (olecranon to tip of long finger) were also used.

Four fingerwidths = 1 palm

Four palms = 1 foot = 36 barleycorns 'taken from the middle of the ear'

Six palms = 1 cubit

The yard, the distance from the tip of King Edgar's nose to the end of his middle finger on his outstretched arm, was legalized as 'our yard of Winchester' by Henry I, the son of William the Conqueror. The mile was derived from the Roman military system, 1000 (*mille*) legionnaires' strides. This was actually 1618 yards but the imperial standard became 1760 yards to the mile (Ritchie-Calder, 1970).

The metre and kilogram were originally chosen by a French committee in 1791. The metre was arbitrarily taken as one-40 000 000th of the circumference of the earth. A standard length was set with which other standard lengths could be compared. In the late 19th century it was a platinum-iridium bar but since 1960 the wavelength of the orange-red line in the spectrum of krypton 86 (an isotope of an inert gas) has been used as the standard.

The kilogram is defined by a platinum-iridium mass of that weight used to check other standards by comparisons made on very sensitive balances. One thousandth of a kilogram, 1 gram, is almost exactly the mass of 1 cubic centimetre of water at 4°C.

The basic unit of time is the second, which is one-sixtieth of a minute and one-3600th of an hour which, unfortunately, are not decimal relationships. The oscillations of atomic systems are currently used to give accurate time measurement.

Table 1.1 gives a list of International System (SI) of Units divided into the seven basic or fundamental units and most of the derived units. The derivation of force has been considered above and many of the others will be discussed in subsequent chapters.

It might be considered that a fundamental unit would be that of momentum but curiously no unit is assigned to this quantity. As already described, the rate of change of momentum which is known as force is measured in newtons ($N = kg\ m\ s^{-2}$). Similarly, velocity is a quantity derived from arbitrary units of length and time, given in metres per

Table 1.1 Some International System (SI) units

Quantity	Unit	Symbols
Fundamental units		
Length (distance)	metre	m
Mass	kilogram	kg
Time	second	s
Other basic units		
Luminous intensity	candela	cd
Amount of substance	mole	mol
Temperature	kelvin	K
Electric current	ampere	A
Some derived units		
Electric charge	coulomb	C
Electric potential	volt	V
Electric resistance	ohm	Ω
Electric capacitance	farad	F
Inductance	henry	H
Magnetic flux density	tesla	T
Force	newton	N
Energy	joule	J
Power	watt	W
Frequency	hertz	Hz
Temperature	Celsius	°C
Pressure	pascal	Pa

second without a special unit. There is, however, a natural unit of velocity – the speed of light, the maximum velocity of radiation, which is approximately 300 million m s^{-1} and called c. In some scientific work velocities are expressed as a fraction of c. (See Chapter 8 for discussion of electromagnetic radiations.)

There are several ways of dealing with large or difficult numbers and units. Firstly, in the interests of simplicity, a recognized abbreviation is used for the units (Table 1.1). Secondly, recognized universal prefixes are used to indicate the magnitude of the unit, e.g. calling one-100th m 1 centimetre (Table 1.2). A way of dealing with numbers, especially large numbers, is known as the power of 10 notation, in which $10^2 = 100$, $10^3 = 1000$, and so on. This is fully described in Appendix A.

The level of accuracy of a measurement is of considerable importance but varies according to the circumstance. In buying a piece of cloth, as in the example above, specifying the length to micrometre level would be absurd and impossible but such precision may well be required for a metal part in an aircraft instrument. Further, the absolute size of what is being measured may influence the level of accuracy required. For instance, a half-kilogram error in a weighing machine may be perfectly reasonable for weighing 50 kg adults, giving a 1% error, but unacceptable for weighing 3.5 kg babies, for whom it would give a 14% error. In some circumstances the accuracy may be more fully expressed by giving the average or mean of several measurements and the extent to which

Table 1.2 Prefixes for SI units

Prefix	Abbreviation	Multiplied factor	Function
tera	T	1 000 000 000 000	10^{12}
giga	G	1 000 000 000	10^{9}
mega	M	1 000 000	10^{6}
kilo	k	1000	10^{3}
hecto	h	100	10^{2}
deca	da	10	10^{1}
deci	d	0.1	10^{-1}
centi	c	0.01	10^{-2}
milli	m	0.001	10^{-3}
micro	μ	0.000 001	10^{-6}
nano	n	0.000 000 001	10^{-9}
pico	p	0.000 000 000 001	10^{-12}
femto	f	0.000 000 000 000 001	10^{-15}
atto	a	0.000 000 000 000 000 001	10^{-18}

they vary from this mean. The word *validity*, when used of a measurement, refers to how closely the measurement conforms to the real dimension being measured. *Reliability* refers in this context to how closely different measurements of the same dimension agree with one another. Notice that in many situations reasonable reliability is required rather than validity. For example, in recording day-to-day changes in weight in the 3.5 kg baby considered above, what may be of importance is the *change* in weight. Providing, therefore, that the machine gives a repeatable reading from day to day and the same machine can always be used, the absolute weight may not be of consequence.

For the rest of this book units will be given by their proper abbreviations according to the SI system (Tables 1.1 and 1.2).

ENERGY

To describe energy it is first necessary to consider energy changes, converting one form of energy to another. For example, electrical energy is converted to heat and light energy in a radiant heat lamp. A list of common kinds of energy with examples of the forms in which they are often manifest is given below:

Heat energy	Heat from the sun or a heat element.
Mechanical energy	Gravitational energy: an object having energy due to its height, say a book on a shelf.
	Energy due to motion, say a book falling from the shelf.
	Energy due to strain or distortion, say the stretch of a spring.
Chemical energy	Energy stored in the chemical bonds of coal, petrol or glucose.
Electrical energy	Motion of charges, an electric current, magnetic attraction.
Radiation energy	Visible light, radio waves.
Energy of matter	Atomic fission, as in nuclear power generation.

Notice that, in the examples, what is being looked at is the conversion or transformation of one form of energy into another or into several others. In other words, energy is only evident during an energy change (see Fig. 4.38).

CONSERVATION OF ENERGY

It is important to understand that there is no beginning and no end to the processes of energy conversion. Energy can neither be created nor destroyed, only changed in form. The total amount of energy of a physical system always remains the same, no matter what energy conversions take place. This is the principle of the conservation of energy. Energy from nuclear power sources may seem to violate this principle in that energy is apparently created from nothing but, in fact, the energy is produced from that stored in the atomic structure of matter; a very small quantity of uranium is ultimately used up.

KINETIC AND POTENTIAL ENERGY

It can be seen that there are two general categories of energy. The one might be described as the energy of motion. It is called kinetic energy. The other is stored energy, referred to as potential energy. In the former, kinetic energy is seen in the movement of solid objects – the book falling from the shelf – or the motion of electric charges as an electric current. Examples of potential energy would be the position of the book on the shelf or the chemical bonding of glucose molecules which release energy in the body when broken down. In some circumstances kinetic energy can be changed to potential energy and back again in a regular manner, forming an oscillating system. A swinging pendulum is an example of such a system in which the motion (kinetic energy) of the pendulum is converted to height (potential energy) at the extreme of each swing and back again. Similar mechanical oscillating systems are seen as vibration (see Chapter 3) and as electrical oscillating systems in which a static electric charge changes to a moving electron flow and back again (see Chapter 4).

MEASUREMENT OF ENERGY

Perhaps the simplest way to imagine energy is to consider the motion of a recognizable lump of matter. It has been noted above that if an object of 1 kg mass is given an acceleration of 1 m s^{-2} it experiences a force of 1 N. If such a force acts through a distance of 1 m it provides a *joule* of energy. So the unit of energy is the newton-metre or joule (J) (Table 1.1). In some contexts energy is referred to as *work* and so the unit of work is therefore also the joule. If a body possesses energy it has the ability to do work.

The joule is quite a small unit. Very approximately it is the amount of energy needed to lift this book 30 cm off a desk or the energy needed to stretch a 10 cm elastic band to three times its length. However, the joule describes the amount of energy and says nothing about the *rate of energy conversion.* In many situations it is this rate of energy change, the number of joules per second, that is of practical importance. Such a measure is called *power* and the units are called watts, so that:

$$1 \text{ watt} = 1 \text{ joule per second}$$

Thus the wattage of a piece of electrical apparatus describes how fast the electrical energy is converted to some other form of energy. Suppose a 250 W infrared lamp is applied for a 20 minute treatment, then electrical energy is being converted to heat at a rate of 250 J s^{-1}, so that during the treatment $250 \times (20 \times 60) = 300\,000$ J or 300 kJ of heat energy is produced. Similarly, an ultrasound treatment applying 0.25 W cm^{-2} for 5 minutes would lead to the application of 75 J cm^{-2} (0.25×300). In neither case would the whole of the energy produced be absorbed into the tissues of the patient.

At other times and in other contexts energy has been described by a variety of units. It can be expressed in watts with the time specified, thus the watt-hour (= 3600 J) and the kilowatt-hour (= 3.6×10^6 J). Other units of energy include the erg (10^{-7} J), which is used in some scientific contexts, the calorie (4.18 J) used in connection with heat energy and the Calorie (4.18×10^3 J) used in nutrition. This rather confusing situation of two units distinguished only by a capital letter has arisen for historical reasons; notice that:

$$1 \text{ Calorie} = 1 \text{ kilocalorie} = 1000 \text{ calories} = 4.18 \text{ kilojoules}$$

The electronvolt is used in describing atomic structure and radiation energy:

$$1 \text{ electronvolt} = 1.6 \times 10^{-9} \text{ J}$$

2. *The structure of matter*

Elements
Atoms
Quantum numbers
States of matter
Gases
Liquids
Solids
Elasticity, plasticity and fracture

Matter is enormously diverse, being made up of many millions of substances, each of which is very evidently different. This tremendous variety is reduced once it is understood that all materials are made of combinations of about 90 naturally occurring substances called elements. (Actually, there are 103 elements but several of the heavier ones are made only in the laboratory.)

ELEMENTS

Elements are made up of identical particles called *atoms*. There are, therefore, some 90 different kinds of atoms. Each of these elements and its constitutent atoms is known by a chemical name and a recognized symbol. Table 2.1 lists some of the elements that are to hand, as it were!

Table 2.1 Some elements occurring in the human body

Atomic number	Name	Symbol	Approximate relative atomic mass
1	Hydrogen	H	1
6	Carbon	C	12
7	Nitrogen	N	14
8	Oxygen	O	16
11	Sodium	Na	23
15	Phosphorus	P	31
16	Sulphur	S	32
17	Chlorine	Cl	35.5
19	Potassium	K	39
20	Calcium	Ca	40

It will be noted that the symbols are simply the initial letters of the element, except for sodium and potassium, because sodium was originally called natrium and potassium kalium.

Apart from small quantities of many other elements – notably iron, magnesium, zinc, cobalt and copper – the human body is entirely made from the 10 elements in Table 2.1.

Thus when several of the same kinds of atoms are linked together the substance is called an element but if different kinds of atoms are united, a vast number of different substances are formed, called *compounds*. The unit due to the combining of two or more atoms is called a *molecule*. Molecules can be made of just two or several hundred atoms. Many of the molecules that make up living tissue are composed of long chains of repeating groups of a few atoms.

The basic building blocks of matter are therefore atoms. Understanding the nature of matter depends on understanding the atom.

ATOMS

Atoms are, of course, incredibly tiny. Even the largest could not possibly be seen directly (although there are techniques that enable the presence of atoms to be made evident in an indirect manner). The approximate diameter of an atom is 3×10^{-10} m, about 0.3 of a nanometre. To put it another way, if 100 million (10^8) atoms were lined up edge to edge they would form a line just about 3 cm long. The mass of each atom is commensurately small and hence rather inconvenient. The mass of the hydrogen atom is approximately 1.66×10^{-24} grams. Hydrogen is the simplest atom so it is easy to describe other atoms in terms of their relative mass, taking hydrogen as 1 unit. Nowadays, however, an isotope of carbon, carbon 12, is used as the accepted standard, its mass being equal to 12 times that of hydrogen. The mass of an atom is called the *atomic weight* of the atom for historical reasons but it is more sensible to call it the *relative atomic mass* (Table 2.1).

Atoms can be considered to be made up of still smaller discrete particles. The size of the atom was described above as being 3×10^{-10} m, but most of this is empty space. This, and much else, was elucidated in 1910 by the New Zealand scientist, Ernest Rutherford. He was able to deduce that most of the mass of the atom was found in a very small central region called the *nucleus*, which is associated with a positive charge. This nucleus has a diameter of about 10^{-14} m, so that the whole atom is some 10 000 times the size of the nucleus. (If the nucleus was scaled up to half a millimetre – like a pinhead – the atom would be the size of a large living room with a diameter of 5 m!) The space surrounding the nucleus is empty of particulate matter but is defined by negative charges moving at high speeds and called *electrons*. The simplest way to describe an atom is, therefore, a small, incredibly dense nucleus made

up of particles, some of which have a positive charge, surrounded by
electrons circling the nucleus at high speeds. This is often described as
the planetary model because of its similarity to the planets orbiting the
sun (Fig. 2.1). However, this is really too simple, because electrons can
neither be thought of as little particles – like tiny moving balls – nor as
having clearly defined pathways around the nucleus. What can be
described is the average position of electrons in relation to the nucleus –
in other words the probability of finding an electron in that position (Fig.
2.2). This pathway, called an *orbital* or *shell*, is often, but not always,
spherical around the central nucleus. When one electron is circling the
nucleus, as occurs in a hydrogen atom, it is easy to imagine its average
motion in the plane of the shell of an egg, with the yolk as the nucleus.
Each orbital has a definite capacity for electrons, after which it is full.
Two electrons can circle in the first orbital or shell, as occurs in the
helium atom. However, if the atom has more than two electrons these
move in spherical shells outside the first. In fact an atom is composed of
a nucleus with electrons circling in a series of concentric shells, each a
greater average distance from the nucleus and containing electrons
which have a specific pattern of movement different from every other
electron.

The nucleus contains two kinds of particles, *protons* which have a
positive charge and *neutrons* which have no charge. Both of these have
significant and similar masses. They are said to have a mass number of
1. The positive charge of the proton exactly balances the negative charge
of the electron so that, as atoms are electrically neutral, the number of

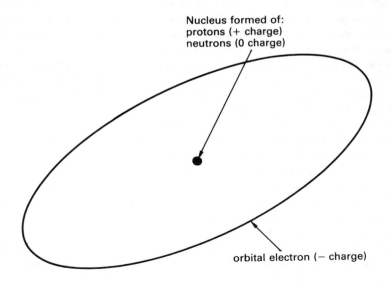

Fig. 2.1 The planetary model of the atom.

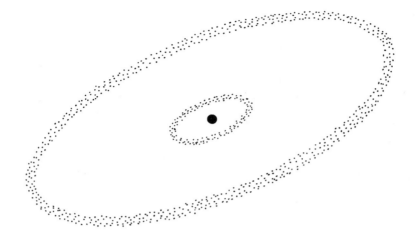

Fig. 2.2 The uncertainty model of the atom. Representation of the spatial probability distribution for electrons in their first two principal quantum numbers.

electrons in its orbital must be the same as the number of protons in its nucleus. Thus in the hydrogen atom there is one proton in the nucleus and one electron in an orbital. Similarly, the helium atom contains two electrons and two protons. Atoms of different elements contain characteristic numbers of protons and electrons for that atom and hence that element. The number of protons (and hence electrons) is given by the *atomic number* (Z) (Table 2.1). The number of proton plus neutrons is given by the mass number (the whole number closest to the relative atomic mass).

If the relative atomic masses are compared with the atomic numbers in Table 2.1 it will be noticed that all elements except hydrogen are twice, or a little more than twice, the mass of their atomic number. This means that the other particles in the nuclei, the neutrons, contribute to the mass and are at least as numerous as the protons. In fact most atoms, especially the heavier atoms, contain more neutrons than protons.

As noted earlier, elements are made up of atoms that are chemically identical to one another. However, they do not necessarily have exactly the same relative atomic masses. This difference is accounted for by atoms that contain different numbers of neutrons. They are called *isotopes*. If an element has different isotopes, the relative atomic mass is not a whole number but reflects the average; see chlorine in Table 2.1. In fact Table 2.1 shows whole numbers for simplicity but many elements are mixtures with very small numbers of isotopes. For instance the relative atomic mass of carbon (C) is strictly 12.01115 because of small quantities of ^{13}C and ^{14}C; however, nearly 99% consists of ^{12}C.

In summary, a simple model of the atom can be described as consisting of a central nucleus containing the mass of the atom and composed of positively charged protons and uncharged neutrons surrounded by a relatively large space in which rapidly moving, negatively charged electrons circulate. This is further summarized in Table 2.2.

In a neutral atom the number of protons is equal to the number of electrons. However, if one or more electrons is removed the particle becomes positively charged and is no longer called an atom. Similarly, if one or more electrons is added it becomes negatively charged. In both cases the atom with a deficiency or excess of electrons is called an *ion* (derived from a Greek word meaning going, in the sense of wandering, since ions in fluids are able to move about).

Table 2.2 Particles of an atom

Particle	Charge	Mass number
Proton	+	1
Neutron	0	1
Electron	−	0

QUANTUM NUMBERS

What has been noted so far is very simplistic. As already mentioned, atoms consist of a number of shells or orbitals numbered from that nearest to the nucleus, called 1, outwards. The shell number is given by the *principal quantum number* (n). The electron shells are made up of one or more subshells which represent different energy levels. These subshells are described by the *second quantum number* (l) and are numbered from 0 upwards. Each subshell can contain a limited number of electrons. These are specified by letters, thus: s can hold 2 electrons, p can hold 6, d 10 and f 14. In a magnetic field these subshells can be further subdivided into orbitals which each hold up to two electrons. This is given by the third or magnetic quantum number (m). The two electrons found in the same orbital turn out to be spinning in opposite directions. This spin is independent of the motion of the electron around the nucleus but is motion of the electron on its own axis. The direction of spin is indicated by the fourth or spin quantum number (s) and can only be one of two possibilities, thought of as either clockwise or anticlockwise. Thus the distribution of electrons in an element, its electronic configuration, and the energy of each electron is different from that of every other electron in the atom (the Pauli exclusion principle) and can be described in terms of the four quantum numbers.

An analogy for the atom is the Roman amphitheatre. The nucleus is the central stage and the electrons are the individual people in the audience. Seats have to be filled from the bottom, nearest the stage, first. The principal quantum numbers (n) represent banks of seats. The subshells (l) are individual rows of seats. The third quantum number (m) is a pair of seats and the fourth (s), rather fancifully, is the way each individual faces (Fig. 2.3).

The lowest possible energy level is the *ground state*: n = 1. The minimum energy required to remove an electron from an atom in its ground state is the *ionization energy*. The energy to remove successive electrons increases. The most easily removed electron is held with the smallest force because it is furthest from the nucleus. The most tightly bound is closest to the nucleus and has a high ionization energy. When atoms interact with one another it is the outermost electrons that will be involved. In fact, the amount of energy holding the outermost electron determines its chemical reactivity; in other words, the way in which the atom with interact with other atoms to form molecules. It is called the *first ionization energy* and varies from atom to atom. It is, of course, a measure of how tightly the outer electron is held to the nucleus. In the same way electronic configuration is linked to chemical behaviour and accounts for the grouping of elements into families in the periodic table. The elements that have a full outer shell of electrons and high first ionization energies tend to be inert. Those that readily lose or gain an electron are especially reactive and have low first ionization energies.

Returning to the amphitheatre analogy – if the exits are only at the top of the theatre, ionization energy is the minimum energy required for a person to reach the exit and first ionization energy is the energy required by the person in the highest occupied level of seats. Thus the nearer the

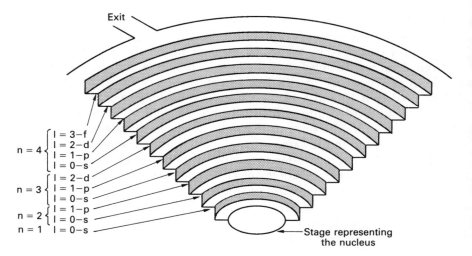

Fig. 2.3 The amphitheatre model of the atom.

stage (the nucleus) the person (electron) is sitting, the more energy is needed to reach the exit.

As mentioned earlier, atoms from different elements join to form compounds; molecules are the smallest possible amount of an element or compound. The formation of the millions of possible molecules, some simply pairs of identical atoms (e.g. H_2), others sheets or chains of different atoms connected together, depends on the union of one atom to another. Some of these, notably the biological molecules, are made almost entirely of carbon, hydrogen and oxygen atoms. They can be quite large, being made up of hundreds of atoms in repeating patterns. This linking or bonding is effected by the interaction of the outer electrons. One way is the sharing of pairs of electrons, *covalent bonding* as it is called. This forms molecules made up of a pair of identical atoms. An opposite way is the union of atoms with different ionization energies. Thus sodium and chlorine can unite to form a salt (sodium chloride). Sodium has a single electron in its outer shell which is easily removed, rendering it a positive ion. Chlorine, on the other hand, lacks an electron to complete its outer shell. If it gains an electron it will become a negative ion. The sodium ion and the chlorine ion will attract each other, the opposite charges serving to hold the atoms into a molecule. This is called *ionic bonding* and in this instance forms crystals of sodium chloride. What is happening in the formation of molecules is rather like the building of very large atoms with several nuclei and shared electrons.

The sharing of electrons in many molecules is not always equal, which causes a charge to exist between the two ends of the molecule, one end being positive relative to the other. This is called a *dipole*. Such polar molecules, the water molecule for example, are readily influenced by an electric field due to their charge.

Changes that are recognized as chemical reactions, such as an explosion, the rusting of iron or the digestion of food, are all rearrangements of atoms at a molecular level in which some bonds are ruptured and new ones are fashioned. In some circumstances, such as the explosion, energy is released. This is the more obvious because the chemical reaction is very rapid. A similar release of mechanical energy may be brought about by changes in chemical bonding during the contraction of a muscle but the rate of energy release occurs more slowly.

From what has been described so far it might be concluded that the protons, neutrons and electrons are fundamental particles, that is, they are not made up of smaller subunits. This may well be a reasonable assertion as far as the electron is concerned, but not for the others. Furthermore, at this subatomic level it becomes impossible to think of particles in terms of solid objects; their energy must be taken into account.

The electron can only have certain definite amounts of energy represented by the quantum numbers. However, the electron can jump between these levels. If an electron changes its orbital to one nearer the nucleus with a lower energy level, energy is released from the atom in the form of a packet or unit of electromagnetic radiation called a *photon*. The frequency and energy of this radiation depend on the energy released by the electron. The photon, however, travels away from the atom and can react with other atoms. It is effectively an energy particle.

By giving the nucleus of the atom a great deal of energy – by colliding protons and neutrons with one another – various other particles have been found. One would expect that protons, having like charges, would repel each other but, in general, protons and neutrons are held together firmly in the atomic nucleus. The force uniting them is called the *strong force*. The particles discovered during these collisions together with particles called *pions* are collectively called *hadrons* (derived from a Greek word meaning bulky because they are relatively large particles). Other particles are formed during interactions in the nucleus. These include the electron and a similar but positively charged particle called a *positron*, as well as an uncharged particle called a *neutrino* and other particles called *muons*. These are collectively labelled *leptons*. Table 2.3 summarizes various particles.

Table 2.3   Hadrons, leptons and the photon

Hadrons		*Leptons*		*The photon*
Proton	+	Positron	+	A packet of electromagnetic energy
Pion	+			
	−	Electron	−	
	0			
Neutron	0	Neutrino	0	
		Muon	+	
			−	

None of this invalidates the model of the atom based on protons, neutrons and electrons but it indicates that the interaction of energy and matter is the basis of structure at a subatomic level. In fact, the laws of conservation of mass and conservation of energy become merged at this level and can be represented by the well-known equation

$$E = mc^2$$

(energy = mass × velocity of electromagnetic radiation squared)

Thus an increase in mass is always associated with a decrease in energy and an increase in energy with a decrease in mass. This is what happens in the sun or in the nuclear reactor of a power station where nuclear fission converts small amounts of matter into large amounts of energy (see Chapter 7).

STATES OF MATTER

So far the nature of the microstructure of matter has been considered in terms of atoms and molecules but the characteristics of matter depend on the way these atoms and molecules are held together. Under the temperature and pressure conditions that exist at the surface of the earth, matter can exist as gases, liquids or solids. (In fact, there are some substances that do not fit exactly into one of these categories.) Further, alterations of the temperature and/or pressure can effect a change of state, as will be seen in Chapter 7. It is helpful to consider the three states of matter separately.

GASES

Gases are the most diffuse form of matter, a form in which there is least structure and in which the atoms or molecules are widely separated from one another. The wide separations allow rapid and random movement of the molecules or atoms. This movement, which can be made visible with a microscope under suitable conditions, is called *Brownian motion* (after the Scottish botanist, Robert Brown, 1773–1858). It consists of irregular zigzag motion as each particle collides with others and bounces off again. Particles will also strike the sides of a container holding the gas. These collisions are elastic so that energy may be transferred from molecule to molecule but is not lost. In fact, this energy of motion – kinetic energy – is part of the energy held in the microstructure of all matter which is recognized as heat (Chapter 7). Typical pathways are shown diagrammatically in Figure 2.4. This *kinetic theory* of gases, as it is called, can be summarized by the following points:

1. The distances between the molecules or atoms are enormous compared to the size of the particle itself.
2. The molecules are in constant, rapid, random, straight-line motion.
3. Because of the distances between them, intermolecular forces are negligible except during collisions.

These account for the well-known characteristics of gases, notably the *universal gas law* which describes the interrelationship of the volume, pressure and temperature for any given sample of gas:

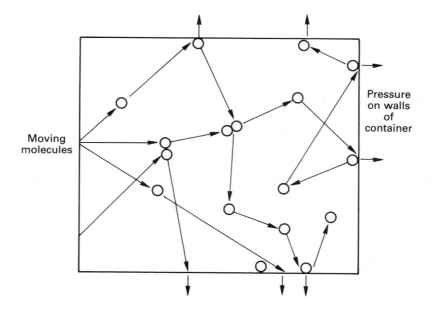

Fig. 2.4 Zigzag pathways of molecules – Brownian motion.

$$\frac{PV}{T} = \text{number of moles of gas} \times \text{the universal gas constant (R)*}$$

where P = pressure, V = volume and T = temperature.

This is combining Boyle's law, the pressure–volume relationship and Charles's law, the volume–temperature relationship. Thus if temperature is constant the pressure of a given volume of gas is inversely proportional to its volume and if pressure is constant the volume of the gas varies with temperature. The pressure is due to the collisions of molecules with the walls of the container. Decreasing the volume, i.e. pushing the molecules closer together, and increasing the amount of movement by heating will both increase the number of collisions on the walls, thus increasing the pressure. Similarly, gases can expand to fill an available volume because the molecules can move further apart. The constant irregular motion also accounts for diffusion which occurs when two gases mix at a rate which depends on their temperature. The universal gas law is not exactly true for gases at very high pressure (at which the volume of the gas molecules themselves becomes significant) and for gases at very low temperature (when the attractive forces between molecules – van der Waals forces, as they are called – slow the molecular motion).

* Gas constant R = 8.314 J K^{-1} mol^{-1}

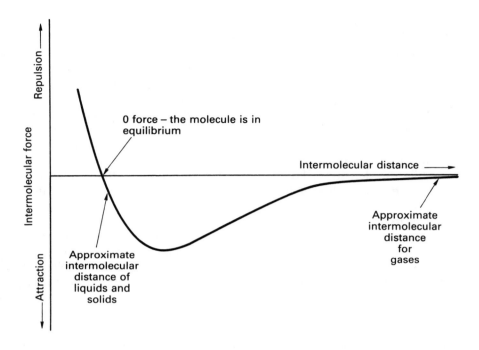

Fig. 2.5 Representation of forces related to intermolecular distance.

As well as forces of attraction between atoms and molecules there must be repulsive forces or they would collapse. The force of attraction between oppositely charged ions is balanced by the electron clouds repelling each other. In a solid the molecules must be more or less in equilibrium. If solids are compressed the forces of repulsion increase and if stretched the forces of attraction increase. A graph of force related to intermolecular distance is represented in Figure 2.5. There is zero force when the molecules are in equilibrium.

In gases the average distance between centres of molecules or atoms is quite large when compared to that of solids and liquids. For a gas at 0°C and 1 atmosphere pressure the average intermolecular distance is about 3.3 nanometres (3.3×10^{-9} m). In liquids and solids the average distance is about a tenth of this. This distance is very close to the point at which the electron clouds of adjacent molecules intermingle sufficiently to produce repulsion forces between molecules (Fig. 2.5).

LIQUIDS

Liquids have a definite volume and density, unlike gases, and take the shape of any container influenced by gravity. They are not rigid but are hard to compress. The density of a liquid is very much the same as that of the equivalent solid but very much greater than that of the gas. Thus a drop of water increases in volume some 1600 times when it is converted to water vapour, i.e. a gas. As noted above, the molecules of liquid are very close together so that the volume of liquid is very much less affected by pressure and temperature changes than that of a gas. Due to the closeness of molecules in liquids the average distance moved during random motion is much less than that of gases. The average distance travelled by a molecule between collisions, called the mean free path, is perhaps tens of nanometres in a gas but only tiny fractions of a nanometre in liquids.

While the density of liquids and close proximity of their molecules make them very like solids, the free random movement of their molecules, albeit over short distances, makes them behave like gases. That is to say, Brownian movement can be observed; they flow and those with similar molecules will diffuse with one another. However, in some instances it is very difficult to make two liquids mix together – oil and water, for example – a fact that is of considerable biological importance. It is also well-recognized that liquids differ greatly from one another in the ease with which they flow – a property called viscosity. This is simply a function of how easily the molecules will move in respect to one another and is unrelated to density. Water and alcohol are well-known liquids of low viscosity while blood, treacle and thick oil have higher viscosity. Unsurprisingly, viscosity falls with increasing temperature since this brings about more molecular movement.

SOLIDS

Solids are clearly different from liquids and gases, which together are known as fluids, because of their rigidity of structure – they do not conform to the shape of their container. They are usually classified into crystalline and amorphous solids. Amorphous solids, such as certain plastics or glass, are in some ways on the boundary between liquids and solids, whereas crystalline solids are formed of regular, repeating arrays of atoms in a three-dimensional structure. The basic unit cell of such structures consists of a number of atoms or molecules in a more or less fixed relationship. There are seven different geometric arrangements to form the unit cell: cubic, like sodium chloride, tetragonal, orthorhombic, rhombohedral, monoclinic, triclinic and hexagonal. These arrays of crystals, called a crystal lattice, are held together by different forces in different substances. Thus:

1. Atoms held together by van der Waals forces or dipole interactions are called *molecular solids*. Their relatively weak, intermolecular forces mean that they melt easily and form soft crystals. Sulphur and solid carbon dioxide (dry ice) are examples. Other molecular solids are

linked by hydrogen bonding. These form rather stronger bonds, forming harder crystals, ice being the most important example.

2. *Covalent bonding* links the atoms in the crystals together in a strong bond with fixed angles between atoms forming very hard, high-melting-point crystals, such as diamonds and quartz. This is rather like fixing the atoms in the crystals into a sort of supermolecule. The atoms share their electrons.

3. *Ionic bonds* are formed when strong electrostatic forces hold the oppositely charged ions together. This is the bonding form of sodium chloride.

4. *Metallic solids* are like a mass of metal ions with positive charges in a sea of electrons. The electrons wander freely through the crystal lattice rather than belonging to a single atom. This accounts for the electrical and thermal conductivity of metals; see Chapters 4 and 7. Of course in an uncharged piece of metal the total number of electrons is equal to the total number of protons. In many metals there is also some covalent bonding, which accounts for the variations of melting points and qualities such as hardness or brittleness of different metals.

In spite of the bonding considered above it would be a mistake to think of atoms and molecules in solids as being stationary. They are in constant vibratory motion which, of course, increases with rising temperature.

Crystal lattices are not pure and structurally perfect. Real solids often have deficiencies in the crystal lattice, called lattice vacancies, and sometimes the lattice is irregular – lattice dislocations. Importantly, the lattice may contain impurity atoms (see Chapter 4).

ELASTICITY, PLASTICITY AND FRACTURE

Figure 2.5 illustrates an important principle for all substances. It is possible to imagine the atoms and molecules as rubber balls, like golf balls, containing magnets. The magnets attract the balls to one another with a force that increases as they get closer but as they press together the rubber is squeezed and tends to push the balls apart. Thus atoms and molecules vibrate about a balanced position between repulsion and attraction.

If a material, say a steel wire, is stretched, it pulls all the atoms apart a little and the attractive force tends to pull them together again – an elastic recoil force. If the stretch is relaxed then the material returns to its original shape. This is called elastic distortion. If the metal is stretched further, some of the crystal lattice array slips past other parts and permanent distortion occurs. This is called plastic distortion or plasticity. Still further stretch may separate the arrays of atoms, breaking the metal; this is called a fracture.

Many of these ideas and concepts will be developed in subsequent chapters. However, it is as well to remember that these descriptions are a very simplified account of the structure of matter, a subject which currently occupies some of the finest brains in the world.

3. *Waves*

Waves can be considered as a way of transferring energy from place to place. There are two principal kinds of waves:

1. *Mechanical waves*, which transmit energy through solids, liquids and gases and involve some changes in the structure or shape of matter.
2. *Electromagnetic waves*, which involve a regular change between electric and magnetic effects.

The former include waves in water and sonic waves, while the latter encompass all the radiations of the electromagnetic spectrum such as radio waves, light and infrared radiations.

MECHANICAL WAVES

Mechanical waves, as mentioned above, include sound and ultrasonic waves, the latter being of frequencies too great to be heard by the human ear, and also waves travelling along ropes and on a water surface. Vibrations are familiar as the movement of solids. For example, if the free end of a wooden or plastic ruler is bent down and released while the other end is held firmly on the edge of the table it will vibrate up and down for a few seconds. Similarly a pendulum, or a child on a swing, is an oscillating or vibrating mass. Such motion is essentially cyclical – a series of events repeated in the same order over and over again, the end of the ruler moving up and down or the child swinging to and fro. If these movements are watched over a length of time the tip of the ruler or swing would trace a wave-like path over time. In fact, if a suitable pen is attached to the end of a pendulum or vibrating string it can be made to trace out the wave motion on a moving sheet of recording paper (Fig. 3.1). This pattern is a sine wave.

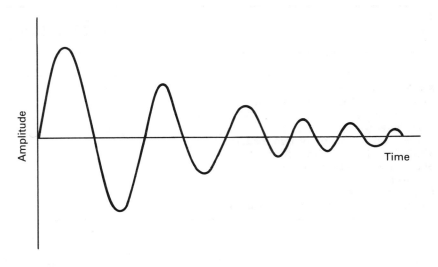

Fig. 3.1 Wave pattern.

Waves are entirely familiar as variations of water surface height – large variations are ocean rollers and tiny ones ripples on a pond. If a coin is dropped into a pool of still water it presses down the water where it enters. Because the water cannot easily be compressed, the coin pushes the water immediately surrounding it upwards into the air. This raised ring of water then falls back due to gravity, pushing up another ring further outwards and so on. Thus the raised ring of water travels further outwards and becomes lower since the energy is now spread over a larger area. At the same time there are more raised water rings being formed. This occurs because, as the first ring falls back, it not only sets up a ring further out but also one inside at the centre where the coin fell. This, acting like another falling coin, sets up another ring of raised water following the first. So a train of waves will pass outwards across the surface of the pool, gradually dying down and eventually leaving the pool completely still once again. Now, although the waves have visibly travelled outwards, it is not the water that has moved towards the edges of the pool. The water has only moved up and down. This is shown by the fact that leaves floating on the surface of the water only bob up and down as the ripples pass under them.

Another familiar example is of a wave moving along a rope or along a coiled spring. If one end of the spring or rope is moved regularly up and down a train of waves is sent along the length of flexible material. A photograph of the flexing spring, showing the waves, is one way of describing waves – illustrating the displacement of all parts at one moment in time – whereas Figure 3.1 is a displacement–time graph for a single point. These are, thus, two ways of illustrating a wavetrain, both of which are shown for a longitudinal wave in Figure 6.1. In both of these the simplest motion is described as a sine wave, which is the plot of the sine of an angle. (At angle 0 the sine is 0 and as the angle increases so the sine increases, at first rapidly and then more slowly, until at an angle of 90° the sine is 1.) The sine wave is simply a relationship between the up-and-down movement and the essentially circular or cyclical motion. It may not be immediately obvious that a leaf, bobbing up and down on the ripples of a pond, is following a cyclical motion. However, consider the changing velocity (speed in a particular direction) of the leaf at each of its different positions (Fig. 3.2). As the leaf bobs downwards the speed increases to a maximum at the midpoint and then decreases until it reaches the trough when the velocity becomes zero. There is then an increasing velocity upwards to reach maximum at the midpoint, followed by slowing up to the crest. Observation shows that the movement stops for an instant at the extreme of each movement, while the leaf is travelling neither upwards nor downwards. Conversely the leaf is travelling fastest while bobbing past the midpoint.

TRANSVERSE AND LONGITUDINAL WAVES

Transverse waves, like waves travelling along ropes, occur where the up-and-down motion is at right angles to the direction of travel of the wave energy.

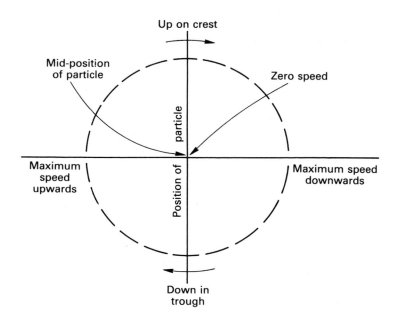

Fig. 3.2 Cyclical motion of a particle involved in wave motion.

Longitudinal waves are waves of compression and rarefaction. If the atoms and molecules of matter are thought of as being pushed together in compression and moving apart in rarefaction, like an accordion, it can be imagined that a wave of compression succeeded by an area of rarefaction can move through matter. It is evident that the individual molecules do not travel but oscillate to and fro. The rectangles in Figure 6.1 are a representation of the particles in matter. If a particular molecule is considered it can be seen that it would move forwards and backwards in the line of travel of the wave. It participates in compression, in which it is closer to other molecules, and rarefactions, in which it is further away from other molecules. The molecule goes through the same sort of cyclical motion as the up-and-down movement of the leaf on the water.

Water waves are neither transverse nor longitudinal. The movement of a particle of water in the path of a wave travelling in very deep water is circular. The plane of the circle will be vertical and the movement of the particle is forwards on the crest and backwards in the trough.

WAVE CHARACTERISTICS

Waves can be described by four characteristics:

1. *Velocity* – the speed of travel of the wave in a particular direction.
2. *Frequency* – the number of cycles of motion, hence waves, per second. The period denotes the time taken for one cycle to occur.
3. *Wavelength* – the distance from a point on the wave until that point recurs, e.g. from crest to crest.
4. *Amplitude* – the magnitude of the wave, i.e. the height of water waves or the pressure in longitudinal waves.

The speed of progression of the waves, the velocity (v), is dependent on both wavelength and frequency. It is found by multiplying the wavelength (λ) by the frequency (f). Thus $v = f\lambda$.

WAVE ENERGY

The quantity of energy that waves can carry will depend on the size and number of waves, i.e. on their amplitude and frequency. Obviously larger waves will involve larger energy exchanges and more of them would carry more energy. When the coin is dropped into the water its kinetic energy is converted into the up-and-down movement of the water. When the front of a wave reaches a certain point the kinetic energy at that point is increased. Work is done in displacing the water from its equilibrium position, so kinetic energy is converted into potential energy. This is then reconverted into kinetic energy and the process is repeated. In this way the wave energy is propagated through the substance.

As waves pass in the medium, energy is gradually lost. This is easily seen on the surface of the pool because the height of the waves, their amplitude, becomes less as they spread out. This spread of the wave over a larger area is one reason for the decreasing amplitude but at the same time energy is being converted at a molecular level, increasing the random molecular motion, i.e. heating, which trivially increases the temperature of the pool. As waves pass through matter the regular wave motion is superimposed on the natural random motion of the molecules. This random motion includes oscillation and rotation of the whole molecule as well as vibratory alterations of molecular shape. In the tissues these latter alterations are of particular consequence in that they represent a significant amount of the random movement of the large protein molecules. Energy is constantly being transferred from molecule to molecule by collisions and as the wave adds motion to the molecules there are more collisions and consequently more random motion which is more heat. Thus the wave gradually loses energy to the medium through which it passes by slightly warming it. This happens, for example, to sound waves in air – the consequent heating is negligibly tiny – and to therapeutic ultrasound waves in the tissues which can cause significant heating because there is so much more energy.

Notice again that the wave energy becomes less, both because it is spread out or scattered and because it is absorbed into the microstructure of the medium through which it passes. Both processes lead to reduction in energy at any point in the medium, called attenuation, but only absorption causes heating or any other effect on the medium itself.

WAVES AT THE JUNCTION OF TWO MEDIA

It is clear that when wave motion tries to pass from one material (medium) in which it travels at a particular velocity to a different material in which it travels at a different velocity there must be a change in the form of the waves if the relationship $v = f\lambda$ is to be maintained. In broad terms, *frequency* remains the same but wavelength and amplitude are affected. It is hardly surprising that the frequency remains constant since the cycles of wave energy arriving at the boundary would cause waves in the second medium to occur at the same time intervals. But the *way* in which wave energy travels in the second medium would be different because of the differing impedance of this second medium.

It has already been noted that the energy carried by a wave is related to the amplitude and frequency of the wave but it also depends on the nature of the medium in which the wave is travelling. Nature of the medium in this context is what is known as *impedance*. Acoustic impedance depends on the density and elasticity of the medium and electric impedance on the electric conductivity and dielectric constant.

It is helpful to consider what happens in terms of a visible slow-moving wave, such as that due to the oscillating movement of a long coil spring, shaken up and down at one end by hand. The medium is the spring itself and its impedance will depend on its density (effectively for a spring, its mass) and on its elasticity. Thus a heavy spring will need more energy to move it and one with little elasticity will also take more energy to stretch it. If the spring is attached to a fixed point, a hook on a wall, say, at one end while the other is shaken up and down, a wave will be seen to travel to the fixed end and then travel back along the spring, i.e. it has been reflected (Fig. 3.3a). Careful inspection will show that, as the crest of the wave reaches the hook, which of course cannot move, the last few coils of the spring are stretched and recoil, pulling the crest down and sending it back along the spring as a trough, i.e. it has been inverted. It is as though the hook in the wall acted to pull the spring downwards. This is an extreme situation because the hook is almost rigid. Consider what happens if the spring is connected to another of different impedance. Suppose the spring is attached to another heavier – greater mass – spring of similar elasticity. When the crest meets the second heavy spring, the energy needed to lift it to the same height as the crest of the first spring would have to be greater because it is of greater mass (Fig. 3.3b). The available energy is insufficient, so it is not raised as much and acts like the fixed hook, stretching the last few coils of the light original spring and thus causing a return inverted wave to be reflected. The result is that part of the energy of the wave continues in the heavy spring while part is reflected back down the original light spring.

Suppose the two springs shown in Figure 3.3b are reversed so that the wave is sent from the heavy to the light spring. At the junction, the wave in the heavy spring lifts the light spring to a higher crest (because it is lighter) which thus transfers a part of the energy back to the heavy spring as a small reflected wave. This time the displacement of the spring is in the same direction, i.e. the wave is not inverted.

It should be noted that, for simplicity, a single wave passing in a spring has been described but the same applies to a series of waves, a wavetrain, if the end of the spring is repeatedly shaken. In these examples a transverse wave has been considered but a longitudinal wave can also be produced in long springs of high-elasticity, 'slinky' springs, but not with the heavy, shorter and stiffer springs used for rehabilitation.

Thus a change in wave velocity occurs at the junction of the two media of differing impedance and a reflected wave occurs. The description given above uses springs to illustrate what happens, but the causes and consequences are the same for all wave motions, for sonic waves or electromagnetic waves. The amount of reflection depends on the difference in impedance between the two media.

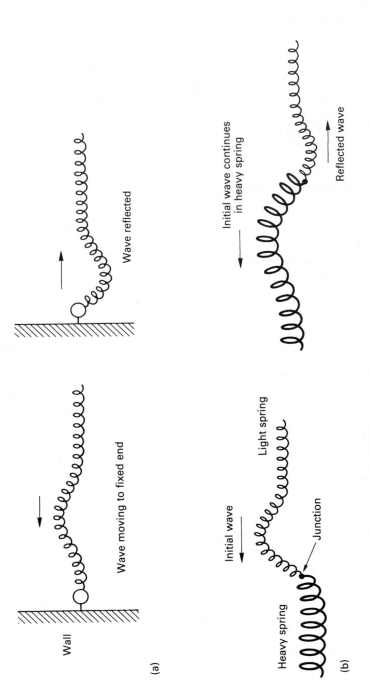

Fig. 3.3 Waves reflected in springs.

A consequence of this change of wave velocity at the junction of the two media is that all the wave energy cannot be transmitted to the second medium so that some of it is turned back or *reflected*. The amount of reflection depends on the difference in impedance between the two media. Thus, if the difference is large there is a large amount of reflection and if there is no difference then no reflection will occur. The ratio of the amplitude of the reflected wave to the amplitude of the original wave (incident wave) is called the reflection coefficient and can be found from the impedances of the two media. If the impedance of the first medium = Z_1 and that of the second = Z_2 then:

$$\text{Reflection coefficient} = \frac{\text{amplitude of reflected wave}}{\text{amplitude of incident wave}} = \frac{Z_1 - Z_2}{Z_1 + Z_2}$$

For sonic waves Z is the acoustic impedance and in the case of electromagnetic waves it is the electrical impedance.

The above description is drawn and modified from Ward (1986), who provides a particularly lucid account of these processes.

It follows from this that to achieve the maximum transmission of wave energy from one material to another, a perfect match in impedance is necessary. However, a reasonably close match will allow the transmission of significant amounts of energy. Of course, in many situations the wave meets a different material which only partly obstructs it. If the wavelength of the wave is larger than the object it meets then the wave tends to bend around it. Thus sound waves pass around most common objects because sound waves have wavelengths of a few metres (for speech and music, lengths from about 0.5 to 5 m would cover most of them). With the higher frequencies of ultrasound, and consequently shorter wavelengths, almost no bending around ordinary objects would occur. Similar comparisons may be made between radio waves and infrared rays; see also Chapter 8.

A second consequence of waves travelling from one medium to another of different impedance is *refraction*. Because of the different velocities the transmitted wave is bent relative to the incident wave. If the velocity is less in the new medium then the wave is bent towards the normal (a line drawn perpendicular to the interface; see Chapter 8).

THE PRINCIPLE OF SUPERPOSITION

Waves can pass through each other without appearing to jostle each other. As they cross they *superpose*; that is, the displacement at any given point is the sum of the individual displacements caused by each wave at that moment. Sometimes the two waves augment each other and sometimes they cancel each other out. This is an important principle and occurs in electrical and mechanical waves. It is made use of in interferential currents (see Chapter 5).

STATIONARY OR STANDING WAVES

If waves are reflected a system can result in which the reflected waves are superposed on the incident waves. At some frequencies a fixed pattern of superposition is produced in which some places have a zero resultant oscillation and some places have a large amplitude of oscillation. This pattern is called a stationary or standing-wave pattern. Places that have zero displacement at all times are called *nodes*. Adjacent nodes are half a wavelength apart. The places with the maximum amplitude of oscillation are called *antinodes* and are also half a wavelength apart.

Standing waves, which arise when a train of waves is partially reflected, can be understood with reference to the transverse waves in long flexible springs, discussed earlier. The reflected wavetrain, meeting the incoming waves, produces a high crest where crests combine at the antinode and no wave (or a wave of no amplitude) where a crest and a trough combine to form a node. The high amplitude, generated at the antinodes, can have important consequences and must be taken into account. Standing waves can be seen using a therapeutic ultrasound machine. The transducer is held in a bowl of water, just below but nearly parallel to the water surface and close to the side of the container. Sonic waves reflected from the side cause standing waves which disturb the surface of the water as visible ripples.

ABSORPTION

When waves are passing through a homogeneous material the same proportion of the available wave energy will be converted to heat (absorbed) in each wave or in each unit of distance that the wave travels. Thus the energy of the wave will become less and less as it is absorbed and so, as it travels further and further, there will be less of it absorbed. Suppose 10% of the energy is absorbed in the first centimetre. In the second centimetre traversed there is only 90% of the original energy, of which 10% is absorbed, which is 9% of the original energy. This leaves 81% of the original energy to pass through the third centimetre and so on. It can be seen that the amount of wave energy will fall exponentially as the wave travels through the medium (Fig. 3.4). This is a very important concept because it describes where the energy is absorbed and hence where it will have an effect. The actual amount of energy absorbed during the passage of waves will depend on the nature of the waves (sonic, electromagnetic etc.), on their wavelength or frequency

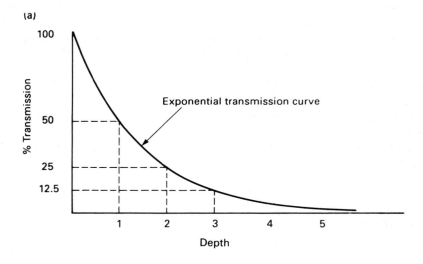

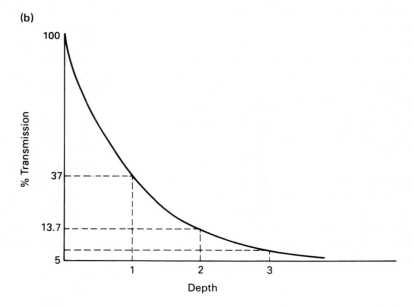

Fig. 3.4 (a) Half-value thickness and (b) penetration depth.

and on the material through which they pass. In order to describe how particular waves are absorbed in particular media it is necessary to describe this exponential change (Fig. 3.4). Since there is no point at which all the energy is absorbed, it is usual to describe the particular curve by a distance at which half the original energy is absorbed, called the *half-value depth*, or by a distance at which some 63% of the energy is absorbed, called the *penetration depth*. This latter is deduced with the use of the mathematical constant e (e is not a whole number; it is 2.718 to four significant figures). The penetration depth is the distance over which the intensity is reduced by a factor of $\frac{1}{e}$ which is 0.37, in other words the distance over which the intensity falls to 37% of its original value. There is clearly a relationship between the half-value and penetration depth. The half-value depth is approximately 0.7 of the penetration depth and the penetration depth is about 1.44 times the half-value depth.

For further discussion on particular types of waves see Chapters 6, 7 and 8.

ELECTROMAGNETIC WAVES

What has been noted about mechanical waves applies to electromagnetic waves in so far as both are wave motion. Electromagnetic waves are different in two important respects. They are variations in electric and magnetic fields and are not due to the positiion of molecules; that is, they are variations in electric and magnetic forces as opposed to variations in mechanical forces. Secondly, they are always transverse waves. They are familiar as radio waves, microwaves, infrared, visible and ultraviolet radiations and also include X-rays, gamma and cosmic radiations (see Chapter 8 and Table 8.1).

Since electromagnetic waves do not involve the movement of particles of matter they travel in empty space. Hence, stars are visible on earth at vast distances because the light they give off is able to travel through the intervening 'nothingness'. This passage through space occurs at a constant velocity of approximately 3×10 m s^{-1} (300 million m s^{-1} or 300 000 km s^{-1}) for *all* electromagnetic waves. Since this velocity results from the product of wavelength and frequency ($v = f\lambda$), it follows that the wavelength varies inversely with the frequency; that is, the greater the wavelength the less the frequency. This means that the differences between electromagnetic waves, visible, infrared etc. are simply due to differences of frequency and wavelength. If the frequency is known the wavelength can be easily calculated from $v = f\lambda$ and vice versa. It is reasonable to describe electromagnetic waves in terms of either but custom and fashion may dictate the preferential use of one. Radio waves were formally given as wavelengths – hence the use of medium-wave band or short-wave band. Nowadays radio stations identify the frequency (see Table 8.3).

Whereas mechanical waves can be seen, electric and magnetic forces have to be imagined. Figure 8.1 is a visual representation of these waves. Notice that the variation in amplitude is sinusoidal; also that the electric and magnetic fields are at right angles to one another and both are transverse to the direction of travel.

To initiate mechanical waves it is necessary to cause some particle/ molecular motion – the coin pressing down on the surface of the water, a drum skin struck to cause compression of air molecules or whatever. But electromagnetic radiations are due to the acceleration of electric charges, that is the motion of charges being changed. This can occur in many ways such as oscillating electric currents or heating (see Chapter 8).

ABSORPTION OF ELECTROMAGNETIC WAVES

While these radiations pass in empty space there is nothing for them to affect but when they pass through matter they will interact with the electromagnetic fields of the molecules and atoms of the material. This will lead to interference with the passage of electromagnetic radiations through matter. There is consequent reduction in velocity and some energy is transferred to the material in which they are passing. The effect of the rapidly varying electric field would be to try to move ions to and fro, to cause alternating rotation of dipoles and to provoke the distortion of the electron clouds of atoms and molecules (see Chapter 8). Such increased motion will result in more atomic and molecular colli-sions adding to the random motion, hence heating. As with sonic waves, electromagnetic waves may provoke specific effects other than heating when they are absorbed, the best known being chemical changes such as photosynthesis or the blackening of photographic film by changing silver bromide molecules.

As the electromagnetic energy is absorbed by the medium through which it is passing, the pattern of absorption/penetration would be exponential in a homogeneous medium. The concepts of penetration depth and half-value depth have already been described. However, the extent of absorption varies enormously for different materials and dif-ferent types of radiation. Thus radio waves will pass almost unhindered through wood, brick, human tissues and many other materials but short ultraviolet radiations are strongly absorbed in air (see Chapter 8).

4. *Electricity*

Electrical phenomena are universal and have been since the universe was formed. However the rapid increase in understanding and the production of electrical machines in the past 150 years have led to a plethora of electrical devices surrounding us and in some ways control-ling our lives. Understanding of such phenomena is therefore central to life in the 20th century.

In Chapter 2 some consideration was given to the forces that maintain the structure of matter. The attraction of the proton for the electron is fundamental to a comprehension of electrical phenomena. Metals were described as crystalline structures with positively charged nuclei fixed within a crystal lattice and lying in a 'sea' of freely moving electrons. The unidirectional movement of such electrons is called an electric current. In fact a movement of any electric charge forms an electric current. Currents are often said to flow but, to be strictly accurate, currents exist and it is the charges that flow. Further, electric charges are not only electrons but include any charged particles, particularly the positive and negative ions formed as the result of the removal or addition of electrons to the atom (see Chapter 2).

In circumstances in which electrons are free to move from atom to atom in a crystalline solid, such as a metal, the current is called a *conduction current* and materials in which such flow readily occurs are called *conductors*. If electrons are added at one end of a long conductor, such as metal wire, and removed at the other end then there is a total shift of electrons along the wire (Fig. 4.1). Notice that what occurs is not electrons travelling down the wire – like a train travelling from one station to another – but the shift of electrons which are already part of the wire. It is like a long tube already full of coloured snooker balls. The addition of a red ball at one end causes the ejection of a ball at the other end of whatever colour happens to be nearest to the end (Fig. 4.2). Of course in the real situation there are many millions of electrons so that the addition of a few at one end leads to the instantaneous displacement of some, but not all, of the electrons along the length of the wire. This conduction current is often described as being analogous to the flow of water in a pipe, the pipe already being full of water.

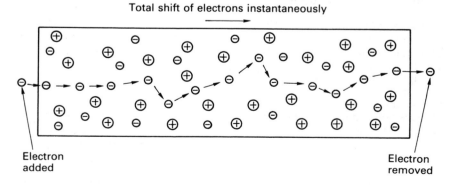

Fig. 4.1   To illustrate the movement of charges or flow of electrons in a conductor – a conduction current.

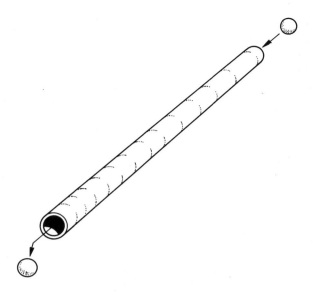

Fig. 4.2 A tube full of snooker balls.

CONDUCTORS AND INSULATORS

Unsurprisingly, the ease with which electrons can move through conductors depends on the nature of the crystalline structure and is not the same for all materials. In fact the nature of the substance dictates the rate of current flow through it – a quality of the material called *resistance*; see below. Materials vary considerably in the ease with which they allow the movement of electrons. In general, metals allow an easy flow, referred to as having a low electrical (ohmic) resistance or a high conductivity. This is, of course, due to the electrons being free to move from atom to atom. Silver is actually the metal with the lowest resistance, closely followed by copper, which is more widely used for obvious reasons! Most pure metals are fairly good conductors but materials in which most of the electrons are more firmly fixed in the crystal lattice are less good but still important conductors because the flow of electrons can be controlled. These are conveniently called *semiconductors*, the best known being silicon. Carbon, germanium and other substances belong to this group, collectively called metalloids, and exhibit resistances several thousand times that of copper. Materials in which there are no free electrons – in which it requires a great deal of energy to move an electron from its parent atom – have very high resistances to electron flow, many million times that of copper. They are called *insulators*. Examples of insulators include most plastics, glass, rubber and dry air. Notice that the distinction between conductors and insulators is a matter of degree; it is a measure of their electrical resistance. Even the best conductors, at room temperature, have some resistance and the best insulators can be made to pass currents under some extreme circumstances.

Electric currents, as already noted, are the movement of charges and so far only electron motion in one direction has been considered. This is called a *direct current* or unidirectional current and is the familiar form of current in battery-operated devices like a torch and in the electrical system of a car. It is, however, just as much a current if the electrons flow first in one direction and then in the opposite direction and is then called an *alternating current*. This is familiar as the mains current which, in the UK, changes direction 100 times in every second, i.e. its frequency is 50 cycles per second or 50 hertz (Hz). The reason for the widespread use of alternating current is principally that its voltage (force) can be easily altered, as discussed later.

Electrons can move very rapidly so that individual electrons can travel from atom to atom very easily. When an alternating current is applied the electrons in Figure 4.1 can be considered as moving to and fro, first from left to right then from right to left. If the alternations are made to occur very much more rapidly, say several millions of times in each second, there is little time for the electrons to move very far and they simply vibrate or oscillate to and fro close to their parent atom. In fact, currents changing direction at these frequencies are often known as *oscillating currents*. If the electrons are only going to oscillate it makes no difference whether they are firmly linked to their parent atom or free to wander – it does not matter whether the material is a conductor or an insulator. Thus high-frequency, oscillating currents can occur in both conductors and insulators; see later discussion on capacitors. Such currents are called *displacement currents* because the electrons are displaced rather than moved.

If the snooker balls of Figure 4.2 are thought of as being attached to the wall of the tube by a short piece of elastic and many balls were added alternately at each end – to simulate an alternating current – it can be seen that the ball movement would be restricted by the elastic. If the alternation were made very rapid so that only one or two balls were being added alternately at each end, it can be understood that the presence of the restricting elastic has little or no effect. Thus displacement currents can exist in insulators in spite of the relative restriction of the movement of their electrons.

So far electric current has been considered to be electron movement. In fluids (that is, liquids and gases) the atoms and the ions are free to move. Thus both positive and negative charges can flow. In fact, a movement of ions constitutes an electric convection current, being the motion of bulk particles in which positive ions pass in one direction towards the negatively charged region and negative ions pass to a positively charged region. This is the way in which most currents pass in the human body. Due to the movement of particles this type of current is called a *convection current*. The chemical changes that can occur when

conduction currents are converted to convection currents will be considered later. Convection currents can be continuously unidirectional
(direct current), alternating or oscillating.

Thus, currents are the movement of charges and are summarized in
Table 4.1

Table 4.1 Different types of current

	Direct current	*Alternating current*	*Oscillating current*
Conduction current (movement of free electrons in conductors and semi-conductors)	e.g. current in torch bulb from battery	Mains current, e.g. in light bulb	In wires of radio circuit
Displacement current (movement of charges in conductors and insulators)			e.g. shortwave diathermy in plastic and tissues
Convection current (movement of charges in fluids – electrons and ions)	e.g. current in tissues for iontophoresis	e.g. current in domestic fluorescent lighting tube	e.g. shortwave diathermy in tissues

DESCRIBING ELECTRICITY

Electric charges and their movements have been described so far
without any reference to quantities. It is absolutely essential to quantify
electric charges and to have a clear understanding of the meaning of the
special terms used.

The basic unit of electric charge is, of course, the electron or ion, but
these are inconveniently small. Thus a unit of quantity, called the
coulomb, is used which is the charge on 6.24×10^{18} electrons. While this
quantity is important, of much more practical use is the rate of
movement of electric charges – the rate of flow – described as the *current
intensity*. This is a quantity of charges passing a point in unit time:

$$1 \text{ ampere (A)} = 1 \text{ coulomb per second (C s}^{-1})$$

Since electric charges transfer energy when they move, it is necessary to measure this electrical force or 'pressure' between two charges, or the energy difference when a charge moves from one position to another. This is known as the *voltage difference* and is measured in *volts* (V). The voltage difference between two points is 1 V if the transfer of 1 C of charge between these two points uses 1 joule (J) of energy. Thus:

$$1 \text{ V} = 1 \text{ J C}^{-1}$$

The rate of converting energy is called power and is measured in *watts* (see Chapter 1), that is to say in joules per second. If the voltage and rate of flow of charges are multiplied it will give the power in watts because the energy conversion rate will depend on the electrical pressure or electrical force – that is to say, voltage – as well as on the rate of flow of electric charges. Compare the power of water, i.e. its ability to do work, which is equal to the water pressure times its rate of flow. Thus:

$$\text{watts (W)} = \text{volts } (V) \times \text{amperes (A)}$$

$$\text{Power} = \text{force of electrons} \times \text{rate of flow of electrons}$$

For example, the 250 W infrared lamp referred to in Chapter 1 is converting 250 J of electrical energy to heat in every second. At the mains voltage of 240 V it would have a rate of flow of 1.042 A, i.e. 1.042 Cs^{-1}.

Electrical equipment of all kinds comprise devices to convert electrical energy to some other form of energy. In the domestic situation the electricity supplier provides homes with a mains supply of charge at standard voltage (240 V in the UK, 110 V in the USA). The user attaches an electrical device which is made to use charge at a certain rate – the wattage. A 60 W electric light bulb will be converting 60 J of electrical energy every second into light and heat. Thus the filament of the bulb must be made to allow a flow of approximately 0.25 A. Similarly a piece of apparatus that is used to convert larger amounts of energy – say an electric fire – would be made to allow larger currents and thus have a higher wattage. For example a 1000 W (1 kW) electric heater would have a rate of flow of over 4 A. It is evident that the total amount of energy will be a function of both the wattage and the length of time. Thus the electrical energy delivered to the domestic consumer is measured in kilowatt hours. Thus it is the energy that is being paid for.

So far the behaviour of the flow of charges has been considered but this is regulated by features of the path of the flow, that is the conductors involved in the electrical circuit. To start with it is simplest to consider a steady, unvarying flow of charges in one direction in a typical conductor, say a metal conductor. As described above, this involves the unidirectional motion of electrons moving fairly easily throughout a crystalline structure, i.e. relatively fixed atoms in a 'sea' of electrons. It takes some energy to cause the electrons to move, that is to superimpose a unidirectional motion on a pre-existing random motion. This opposition to charge flow is called *resistance* and is a characteristic of the circuit or pathway. The longer the pathway the more electrons must be moved. Also it is to be expected that some materials, like metals or metal alloys, will allow electrons to move more easily than others, simply because of their structure. Furthermore if the random motion of electrons and atoms is increased by heating (see Chapter 7) there will be random collisions and hence more energy used in maintaining electron movement. Thus it can be seen that the resistance of a metal conductor will depend directly on its length and its temperature and on the kind of metal concerned. If the choice of pathway for the electron motion is restricted by using a conductor with a small cross-sectional area the resistance is also increased. This quality of the circuit, its resistance, is measured in *ohms* (Ω) and is most easily defined in terms of the current intensity and voltage. Thus a circuit with a resistance of 1 Ω will allow 1 V to drive a current of 1 A. This relationship is central to any consideration of electrical phenomena and is known as *Ohm's law*.

Consider Figure 4.3, in which the source of voltage, known as the electromotive force (e.m.f.), and the pathway that the current follows are shown diagrammatically. In such diagrams the lines represent conducting wires and the boxes represent resistances. (In real circuits all the wires will have some resistance but this is usually negligible compared to the load, so that it can be ignored.) In the circuit of Figure 4.3 applying a voltage across the resistance will lead to a current. Increasing the voltage will increase the current and vice versa. Thus it can be said that for any given resistance the current is directly proportional to the voltage. This is expressed in Figure 4.4, which is a graph of current against voltage for a given circuit. Similarly, if the resistance in Figure 4.3 were to be changed for one of less resistance then the same voltage would drive a larger current through the circuit.

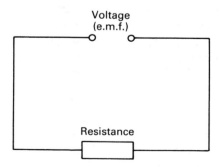

Fig. 4.3 A simple circuit diagram.

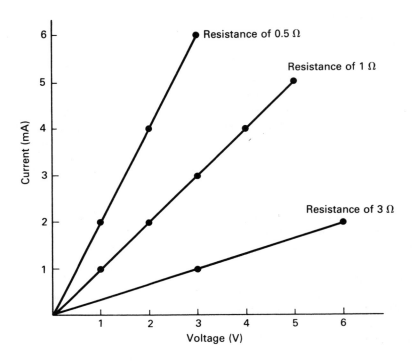

Fig. 4.4 A graph to show Ohm's law.

Thus it can be said that the current is inversely proportional to the resistance. The relationship can be expressed as:

$$\text{Current} = \frac{\text{voltage}}{\text{resistance}} \text{ or } I = \frac{V}{R}$$

The units are often denoted by their initial letters, A for amps and V for volts, but it is customary to use the Greek letter omega (Ω) for ohms. It is obvious that knowing the numerical value of any two of the units allows calculation of the other. Thus if in Figure 4.3 the resistance were 60 Ω and the mains voltage of 240 V were applied then a current of 4 A would result, which is approximately appropriate for the 1000 W heater considered above.

The concept of Ohm's law is absolutely fundamental to understanding electricity and electronics. It is, in fact, the electrical equivalent of a common experience and is only less familiar because the movement of charges is an invisible process. When a water tap is turned on, a flow of water results driven by the head of pressure, often due to the height of water contained in a tank in the roof. If the tap is turned off partly, the flow is reduced to a trickle because of increased resistance to the flow in the tap. Sometimes such a system is driven by a pump instead of the force of gravity. In either case this is the driving force equivalent to the voltage in electrical systems. If this driving force is increased, say by raising the height of the tank or force of the pump, it will increase the rate of flow of water, equivalent to the flow of charges or current intensity in amps. Similarly, when the tap is partly turned off it decreases the flow of water in the same way as increasing the resistance affects the current intensity. Such a system operates in the human body to regulate the flow of blood to the tissues. The heart pumps blood whose flow to various tissues and organs is regulated by varying the diameter of arterioles. The rate of flow can thus be made greater by increasing the force and rate of the heartbeat – the driving force or e.m.f. – as well as by dilating the arterioles to lower the resistance or peripheral resistance, as it is called here. What Ohm's law is stating is that the voltage across some component or part of the circuit is proportional to the current and the proportionality factor is the resistance, shown as the slope or gradient of the graph in Figure 4.4. From Figure 4.4 it is evident that the straight line through the origin describes an unchanging resistance, hence it is called a linear system. There are situations in which a component in the circuit changes its resistance as the current passes and thus exhibits non-linear behaviour. Such behaviour occurs with semiconductor devices, such as transistors, in the circuit and also with such simple devices as tungsten filament light bulbs. The reason for this latter effect has already been noted when the dependence of resistance on temperature was considered. At the instant of switching on, a large surge of current passes in the cold filament (it will be many times the 0.25 A in the 60 W bulb considered above) but as the filament rapidly heats up, its resistance increases to something like 15 times that of the cold filament.

Placing resistances end to end, as in Figure 4.5a, is equivalent to a single longer resistance. As noted earlier, resistance is proportional to length so by adding the three 2 Ω resistances together a resistance of 6 Ω results.

$$\text{Total } R = R_1 + R_2 + R_3$$

The voltage drop across each is proportional to the resistance (R) of each. As, in this case, there are three equal resistances, the voltage drop across each is a third of the whole (12 V) i.e. 4 V across each.

The same reasoning applies to Figure 4.5b in which the resistances have different values. The total resistance is determined by adding the individual resistances but this time there is a different voltage drop across each. The voltage drop across and the current through each resistance are:

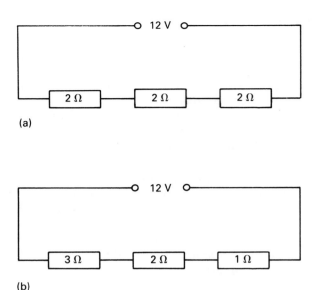

Fig. 4.5 Resistances in series.

				Total
Resistance	3 Ω	2 Ω	1 Ω	6 Ω
Voltage	6 V	4 V	2 V	12 V
Current	2 A	2 A	2 A	2 A

Note again that the current is the same in each apart of the circuit. Resistances arranged in this manner are said to be *in series*.

Figure 4.6a illustrates a set of resistances arranged in parallel. In this case the same voltage is applied to all three so it is the current which varies in each:

$$I \alpha \frac{1}{R}, \text{ in this case 6 A.}$$

Now if 6 A passes through each parallel pathway the total current would be 18 A. This is not surprising since providing parallel pathways is like making a single ̲sistance with a larger cross-section. The total resistance is found by the formula:

$$\frac{1}{R} = \frac{1}{R_1} + \frac{1}{R_2} + \frac{1}{R_3}$$

Thus:

$$\frac{1}{2} + \frac{1}{2} + \frac{1}{2} = \frac{3}{2} = \frac{1}{\text{Total resistance}} = 0.667 \ \Omega$$

The total current through all three resistances will thus be:

$$\frac{12 \ V}{0.667 \ \Omega} = 18 \ A$$

as already noted. Figure 4.6b illustrates the situation with three resistances of different values in parallel. The voltage drops across and the current through each resistance are:

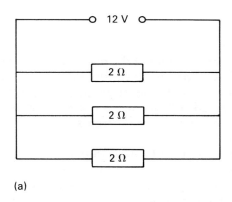

(a)

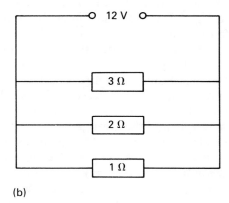

(b)

Fig. 4.6   Resistances in parallel.

				Total
Resistance	3 Ω	2 Ω	1 Ω	0.5454 Ω
Voltage	12 V	12 V	12 V	12 V
Current	4 A	6 A	12 A	22 A

Adding the resistances:

$$\frac{1}{3} + \frac{1}{2} + \frac{1}{1} = \frac{11}{6} = \frac{1}{\text{Total } R} = 0.5454 \ \Omega$$

Total current will therefore be:

$$\frac{12 \ V}{0.5454 \ \Omega} = 22 \ A$$

RESISTANCES

It is plain that the arrangement of the resistances and their magnitude in a circuit will partly determine the current flow. In fact many electric and electronic circuits consist of resistances with other components. In this context they are called *resistors* and are often manufactured to have a particular resistance. For the small currents of electronic circuits they can be made quite small.

These are fixed resistors but a most important function of a resistor in an electrical or electronic circuit is to regulate the current and for this a variable resistor is used. In principle this simply involves being able to alter the length of the resistance wire in the circuit. By far the most familiar types are those that consist of a sliding contact moving over the resistance wire as a knob is turned, thus putting more or less resistance into the circuit (see Fig. 4.7, which also shows the symbol for a variable resistance). It is fairly easy to see how a variable resistance might be used in series in a circuit to regulate the current but a much more appropriate method is by varying the resistance of a parallel circuit. These arrangements are shown in Figure 4.8. The current in the load circuit depends on how much resistance is included. The arrangement of Figure 4.8a in which the variable resistance is in series with the load is sometimes called a *rheostat* and the arrangement of Figure 4.8b, which is widely used to control currents applied to patients, is sometimes called a *potential divider*, because it divides the potential difference between the circuits, or a *shunt rheostat*, shunt referring to a parallel circuit. This latter arrangement has the important advantage that the current in the load circuit can be made zero. If the reasons for this important consequence are not obvious, see Appendix B.

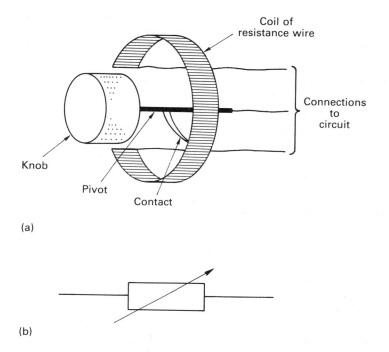

Fig. 4.7 (a) A variable resistor and (b) its symbol.

Table 4.2 Summary of electrical units

Name	Meaning	Symbol
Coulomb	Number of electrons – quantity	C
Ampere	Flow of charges – C s^{-1}	A
Volt	Force or electrical pressure – J C^{-1}	V
Watt	Power – J s^{-1}	W
Units relating to the conducting pathway		
Ohm	Difficulty of charge flow – resistance – V A^{-1}	Ω
Farad	Capacitance – C V^{-1}	F
Henry	Inductance – rate of current per volt	H

Conductance is the reciprocal of resistance; it is the ease with which currents flow. The unit of conductance is the mho.

So far the units shown in Table 4.2, pertaining to the movement of charges, have been considered. All these have been given as the basic

Physical Principles Explained

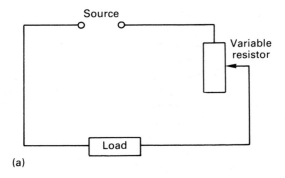

(a)

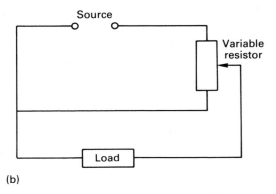

(b)

Fig. 4.8 A variable resistor (a) in series with load – a rheostat; (b) in parallel with load – a potential divider.

unit but in practical situations both larger and smaller units are of interest. Thus the amp is quite a large unit and most currents applied for therapy are of the order of milliamps (mA) or microamps (μA). Large voltages are found in some situations such as the national grid where voltages of hundreds of kilovolts are applied. Kilowatt hours have already been noted (see Chapter 1 and Appendix A).

MEASUREMENT

Meters or measuring devices are often placed in electric circuits to measure these quantities. Ammeters, a contraction of amperemeters, are used to measure the rate of charge flow and to do so properly must be in the path of all the current, hence in series in the circuit. (Milliammeters are used to measure small currents.) Since this instrument is a part of the circuit it must be made to have a very low resistance itself so that it does not significantly alter the current it is measuring. A voltmeter is used to measure the voltage drop across a particular part of the circuit, hence it is in parallel. Conversely it needs to have a very high resistance so that it does not allow any significant flow of current which would alter the voltage it is measuring. The way in which many of these instruments work is described in principle later in the section on electromagnetism. The resistance of a circuit can also be measured by an ohmmeter. A simple one contains a small battery and the current through the circuit being investigated is measured and the resistance given on the meter by Ohm's law.

DIRECTION OF CHARGE FLOW

So far consideration has only been given to unidirectional currents which have been described in terms of charge (electron) movement. Now electrons travel from regions of high negative potential, i.e. excess electrons, to regions with lower negative potential, i.e. with few electrons, described as positively charged. Therefore electrons travel from negative to positive. However, for historical reasons current is said to flow from positive to negative. This may lead to confusion in reading descriptions but it causes no practical difficulty; for unidirectional current circuits it is usually simpler to think of electron flow. As will be seen, many situations involve alternating currents for which the direction of flow is rapidly alternating so that there is no difficulty.

STATIC CHARGES

Before proceeding to consider the effects that electric currents can produce, it is important to think about the effects that electric charges themselves can cause when they are not moving, hence often referred to as *static electricity*. The obvious way in which charges are held in one place is to generate them on the surface of a good insulator so that they cannot flow away. This is a well-known phenomenon which occurs as a result of rubbing two insulators together so that electrons are rubbed off one and on to the other. A homely example is the crackling resulting from pulling nylon clothing over the head in dry conditions or the sparking that can sometimes be produced by rubbing clothing on plastic chairs. If a polythene rod is rubbed with a piece of dry nylon the rod and the nylon can attract little pieces of paper. This occurs because electrons are rubbed off the nylon on to the rod which is therefore now negatively charged and thus attracts and will hold the uncharged pieces of paper. If two polythene rods are hung by threads close together and both charged by rubbing with nylon they will be repelled from one another. This is to be expected since both will be similarly charged and the excess electrons on each rod will give a repulsive force.

Two points need to be noted about this electrostatic force between objects. Firstly, the force will act both through space (a vacuum) and through matter, e.g. air. Secondly, the force will diminish with increasing distance between the two objects. In fact, the force between two charges is inversely proportional to the square of the distance between them. The fact that this effect occurs when the objects are not in contact led to it being called induction – electrostatic induction. This is because one charged object can induce a charge on a second object at a distance. It is important to consider how this can come about. If a negatively charged body is brought close to a conductor, as indicated in Figure 4.9, the electrons in the conductor are repelled from the surface nearest to the charged body. The electrons themselves cannot pass from one body to the other but the force is transmitted and influences electrons in the other body. Thus a negative charge on one body will drive electrons away from the near surface of the other body to leave it positively charged, as shown in Figure 4.9. Similarly, a positive charge on one body would induce a negative charge on the near surface of the other. This effect will occur if there is a vacuum between the objects but will occur with more force if some other insulating material is interposed. This is because the paths of the orbital electrons of the insulating material are distorted by the charges so transmitting the force, even though the electrons of the insulator are firmly held to their parent atoms, as suggested in Figure 4.10. When an insulator is acting in this way it is called a *dielectric*. Some materials will be more effective than others at transmitting the charge, as the electron 'clouds' will distort more easily without breaking down, i.e. without the electrons moving from one atom to another, as discussed below.

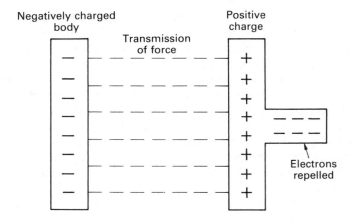

Fig. 4.9 A charged body brought close to a conductor.

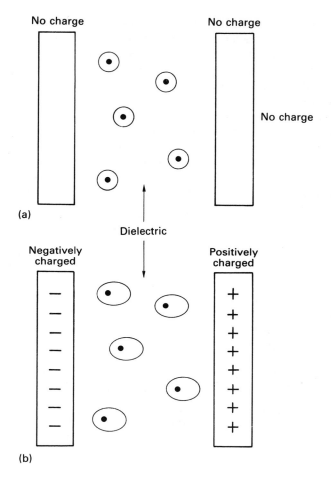

No charge   No charge

No charge

(a)

Dielectric

Negatively charged   Positively charged

(b)

Fig. 4.10 The effect of a dielectric.

CAPACITORS AND CAPACITANCE

Capacitors are devices used to store electric charges, based on the principle of electric induction. They are basically a pair of conductors separated by an insulator which is the dielectric. The amount of charge that can be held will depend on the size of the capacitor, i.e. the area of the opposing plates. It will also depend on the nature of the dielectric measured as the dielectric constant (see below) and the distance separating the conductors, i.e. the thickness of the dielectric. The capacitors for use in electrical devices are made to maximize their capacitance in a conveniently small size. For example, a typical capacitor might be made of two thin sheets of aluminium foil separated by a flexible dielectric such as waxed paper and rolled into a cylinder to give a convenient shape (Fig. 4.11a). Commercially made capacitors can be of many kinds, often described by the material of the dielectric, such as mica capacitors, polyester capacitors and ceramic capacitors. Electrolytic capacitors consist of the two metal plates separated by a layer of metal oxide. These can have a high capacity, at least for low voltages, because the oxide layer is very thin. They are polarized capacitors so that they are fitted in the circuit only one way, unlike the other capacitors where the terminals can be either positive or negative.

Another important kind is the variable capacitor in which the relationship of the plates, and hence the capacitance, can be altered. These are usually made of two sets of rigid parallel metal plates separated by air. Each set is electrically connected to form one conductor and the air is the dielectric (Fig. 4.11b). The symbols used to indicate capacitors in circuit diagrams are also shown in Figure 4.11.

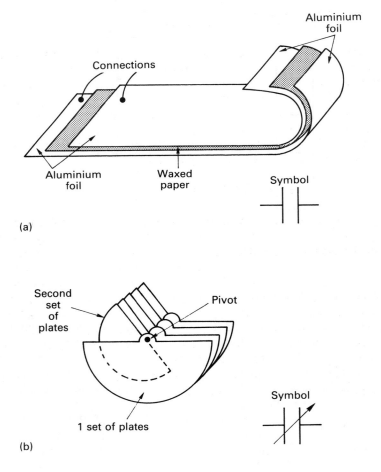

Fig. 4.11 Capacitors: (a) a fixed capacitor and its symbol; (b) a variable capacitor and its symbol.

From what has been noted, it is evident that capacity is a quality of a device depending directly on the effective area of the plates and the type of the dielectric (the dielectric constant or relative permittivity, discussed later) and inversely on the distance between the plates. It is measured in *farads*.*

An analogy can be thought of in terms of the distortion of a rubber diaphragm across a pipe of water. The amount of distortion depends on the surface area of the diaphragm, the type of rubber it is made of and its thickness – the thicker the rubber the less the distortion. It is also evident that for any given capacitor the amount of charge that can be stored will depend on the force applied, that is the voltage across the plates. If a voltage of 1 V stores a charge of 1 C on a capacitor then the capacitor has a capacitance of 1 F, i.e.:

$$\text{Capacity in farads} = \frac{\text{quantity in coulombs}}{\text{voltage in volts}} \text{ or } C = \frac{Q}{V}$$

An analogy can be made with tanks of water. A container with a large base area will need more water, more quantity, to reach the same depth/height and thus have the same pressure (Fig. 4.12). In the same way a capacitor with a large capacitance will need more charge – greater quantity in coulombs – to raise the electrical pressure, in volts, to the same level as a smaller capacitor.

The farad (F) is quite a large unit and most practical capacitors are found to be in microfarads (μF) or picofarads (pF):

$$1 \ \mu F = 10^{-6} \ F \text{ and } 1 \ pF = 10^{-12} \ F$$

It will be recalled from Chapter 2 that atoms and molecules have orbiting electrons – electron 'clouds' – which, being negatively charged, are influenced by an applied electric field. Thus if a negative charge is applied to one side of a block of insulating material and a positive charge to the other the electron clouds of all the atoms and molecules will be distorted in the same direction, away from the negative and towards the positive charge (Fig. 4.10). Each individual atom or molecule has become *polarized*, i.e. it has a negative and positive charge at each end. It will also be recalled from Chapter 2 that some materials, notably water, have molecules which are polar (called dipoles) even without an applied electric field. If the sides of a piece of such material are oppositely charged the dipole molecules will swing round so that they all face in the

---

* Capacitance is $8.84 \times 10^{-12} \times \dfrac{AK}{d}$ farads

where *A* is the effective area of plates in m², K is the dielectric constant and *d* is the thickness of dielectric in m.

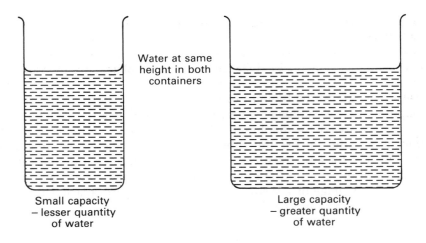

Water at same
height in both
containers

Small capacity
– lesser quantity
of water

Large capacity
– greater quantity
of water

Fig. 4.12   To illustrate capacity.

same direction, negative end to positively charged side and vice versa.
In both circumstances some energy is stored by the charge as an
alteration or distortion of the material of the dielectric. The effect is
generally greater with polar than with non-polar materials. The dielectric constant or relative permittivity of a material is related to its degree
of polarization. It is simply the ratio of the capacitance of a capacitor
with the material between the plates to the capacitance with a vacuum
between the plates. Table 4.3 gives examples of the relative permittivity
of some materials.

It will be noted from Table 4.3 that water has a very high relative
permittivity of 81 because, as noted in Chapter 2, it is markedly polar.
This allows a simple demonstration of the effect of an electric charge on
the flow of charges, something that occurs in a number of electrical
devices such as oscilloscopes or electron microscopes. If a thin stream of
water is allowed to run from a tap and the plastic barrel of a ballpoint
pen which has been rubbed with a piece of nylon is brought close to it,
the stream will be seen to change direction. The water molecules are
sufficiently polar to be influenced by the static electric field.

In this connection the distinction between insulating properties and
dielectric properties may be noted. Insulators offer a very high resistance to the flow of charges but this is independent of the dielectric
properties, which depend on the polarization. Of course, to function as
the dielectric of a practical capacitor the material needs both dielectric
and insulating properties. It would not work as a capacitor if charges
could pass from one plate to the other. This can happen in any material
if the voltage is high enough so that the dielectric strength is often
specified for capacitors.

Table 4.3 The dielectric constants or relative permittivities of various substances

	Material	*Relative permittivity*
Gases	Hydrogen	1.0003
	Air	1.0006
Non-polar substances	Polythene	2.3
	Benzene	2.3
	Beeswax	2.7
	Paper	2.7
	Mica	5.4
Polar substances	Ethyl alcohol	25.8
	Ethanol	28.4
	Water	81.1

If two capacitors are connected in a circuit in parallel it is equivalent to a single capacitor with an increased area of the plates. The total capacitance is simply the sum of the individual capacitances in parallel. If the capacitors are connected in series it is the equivalent of a single capacitor with a thicker dielectric so that the total capacitance will always be less than the sum of the individual capacitors and may be found from:

$$\frac{1}{\text{Total capacitance}} = \frac{1}{C_1} + \frac{1}{C_2} \text{ etc.}$$

CAPACITANCE RESISTANCE CIRCUITS

Having considered the concepts of resistance and capacitance, it seems appropriate at this point to see how they might be combined in a real electric circuit and what effects they might have. It will be recalled that a capacitor can store a charge, i.e. an excess of electrons on one plate and a deficiency on the other. Now suppose the two plates of the charged capacitor are connected to one another through a resistance (Fig. 4.13). Current would occur as electrons flowed from the negative plate of the capacitor to the positive via a movement of electrons in the resistor. The charged capacitor is said to have discharged through the resistor. For any given capacitor the voltage depends on the size of the charge, i.e. the number of electrons, so that as current flows so the voltage driving it falls and, of course, as the voltage falls so the current will diminish (Ohm's law). Consequently, the capacitor will discharge through a resistance with an exponential fall of voltage and current (Fig. 4.13). Clearly the length of time for discharge of the capacitor will depend on the size of the charge (number of electrons) and the size of the current that is allowed to flow. The former will depend on the capacitance and the latter on the resistance. Increasing the capacitance allows more electrons to be stored for a given voltage and increasing the resistance allows a smaller current, hence it will take longer for a large capacitor to discharge and still longer if it is doing so through a larger resistance. The mechanism is analogous to a sand-in-glass egg-timer. The time it takes for the upper compartment to empty of sand depends on the number of sand grains (the charge) and the size of the narrow neck (the resistance; Fig. 4.13). The force of gravity is equivalent to the voltage. Both devices are used for timing. The capacitance–resistance (CR) circuit is used to time electrical pulses in many situations, from those of a few microseconds used therapeutically to those of several seconds or minutes, such as traffic lights. The length of time taken for discharge can be exactly adjusted by altering the size of the capacitor or, more usually, the resistance. Fitting a variable resistance can allow the pulse length to be altered at will and this is how therapeutic sources with pulses of variable length operate. What has been said of the discharge of a

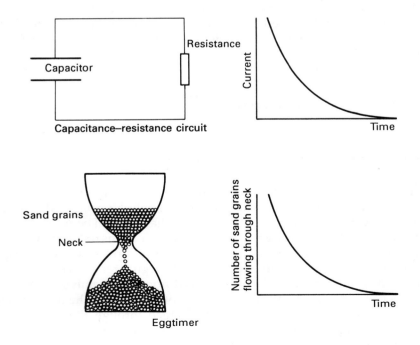

Fig. 4.13 The relationship of rate of flow to time.

capacitor through a resistance applies to the charging of a capacitor. In this case the effective voltage falls as the charge on the capacitor becomes closer to the voltage of the charging source. How such circuits are charged and discharged is considered further on page 121.

The graph shown in Figure 4.13 is a plot of current against time. Graphs that illustrate current (or voltage) changes over time are widely used to describe currents and pulses of current. They are able to illustrate the rate of change, rise or fall, of the current, as in Figures 4.13 and 4.14. They also show the duration of current and, if the current is in the form of discrete pulses, the interval of time between the pulses, hence the frequency (Fig. 4.15). Finally, they describe the direction, i.e. if the current passes first in one direction then in the other (Fig. 4.16). Current passing in one direction is called *direct current*: see, for example, Figures 4.14 and 4.15. When such direct current passes for a significant length of time it is sometimes called continuous direct current or in therapeutic situations constant current, uninterrupted direct current or galvanic current, the last being a rather out-of-date term. When the current passes first in one direction and then in the other it is called alternating (Fig. 4.16) and if the pulses in each direction are equal it is called an evenly alternating current, e.g. Figure 4.16b. In fact, since evenly alternating currents are widely known in the form of the mains current it is usually simply called alternating current. If the frequency of alternations is very much higher than the 50 Hz of mains current it is often called oscillating rather than alternating current.

Suppose a voltage were applied across a resistance such that the voltage rose, continued at a steady pressure for a few seconds and then fell: the current-against-time graph shown in Figure 4.14a would result. However, if a voltage changing in the same way were applied across a capacitor, no steady current would flow – only a small pulse of current at the start while the capacitor was charging and at the end while it was discharging. If the resistance, in the first case, were made larger, the magnitude of the current would be less and vice versa (Ohm's law; see page 41). However, any size capacitor in the second case will completely block the steady flow of current. Now consider what happens with a short pulse of current such as the 1 ms pulse of Figure 4.14c. In this case the rise-and-fall times of the voltage are a much greater proportion of the total time of the pulse, hence so are the charging and discharging currents. With still shorter pulses this effect becomes greater so that when the current starts to fall immediately it stops rising, i.e. there is no steady current, the presence of the (insulating) dielectric becomes inconsequential. Thus very short pulses are transmitted in a circuit with a capacitor just as well as in one without a capacitor. Consider the circuit shown in Figure 4.17 and the current that would flow in the horizontal resistance if a continuous voltage were applied at the source. This time the current has alternative pathways. Its magnitude would depend on the voltage of the source and, by Ohm's law, the sum of the three resistances (because they are in series). If, however, a

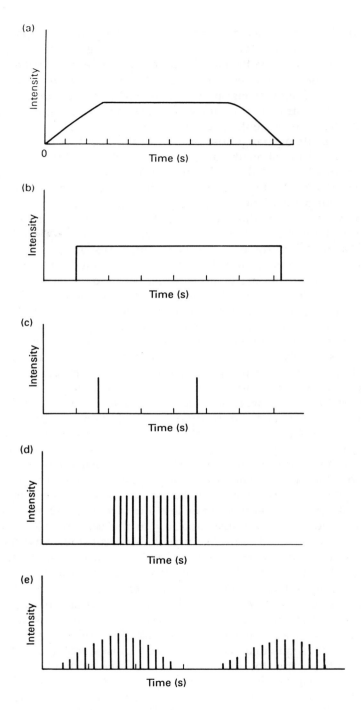

Fig. 4.14 The relationship of time and current: (a) direct current; (b) rapid rise and fall in current; (c) 1 ms pulses; (d) 1 ms pulses repeated every 10 ms; (e) surged current.

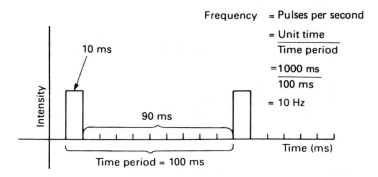

Fig. 4.15 The relationship of frequency to time period.

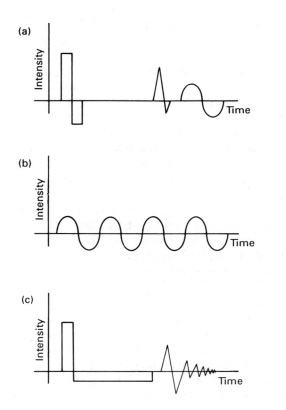

Fig. 4.16 Different forms of biphasic current: (a) discrete pulses; (b) continuous pulses; (c) asymmetrical pulses.

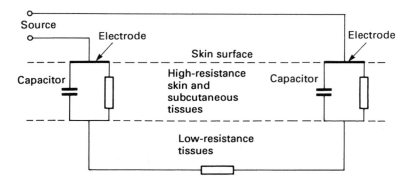

Fig. 4.17 Electrical pathways of current applied to the tissues.

short pulse of the same voltage were applied it would be largely transmitted through the capacitors, thus partly bypassing the vertical resistors. Because of this the magnitude of the short pulse of current through the horizontal resistance would be greater than in the first case because of the lower effective resistance.

If the above is not immediately clear the effect can be illustrated with a water analogy. The capacitor is analogous to a tank of water divided by an elastic diaphragm impervious to water (Fig. 4.18). The two halves of the tank, the storage chambers, are equivalent to the plates of the capacitor. The rubber diaphragm represents the dielectric which does not permit the passage of water from one chamber to the other in the same way that electrons cannot cross the dielectric. In both, a pressure of water or electrons can cause distortion, temporarily storing energy. The narrow, convoluted pipes represent the vertical and horizontal resistance. Now if a bucket of water were removed from the right-hand tank and *slowly* poured into the left-hand tank from the top it would distort the left diaphragm downwards slightly, pressing water out of the lower chamber along the narrow horizontal pipe to the right-hand side. At the same time, over the next few seconds, water would also flow down the vertical resistance pipe, producing a steady left-to-right continuous flow in the horizontal pipe as the flow from the bucket continued. Ultimately the diaphragms would return to their original positions. If the whole bucket of water were poured in at once the vertical pipe could not transmit that amount of water quickly enough but the left-hand diaphragm would be strongly distorted, causing a flow of water in the horizontal pipe and filling of the right-hand lower chamber, with consequent distortion of that diaphragm. In other words, with a slow trickle of water the majority is transmitted through the resistance pipe. With a short-duration charge, or in this case quick emptying of the bucket, the resistance pipes are bypassed and the force is transmitted through the main tank via distortion of the diaphragm and the rate of flow is greater through the horizontal pipe.

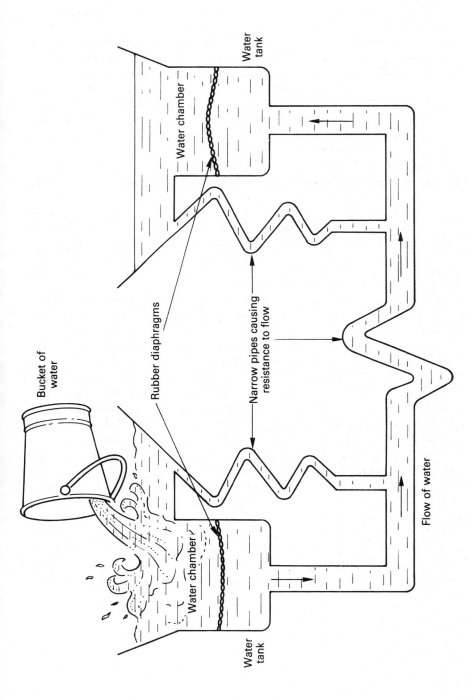

Fig. 4.18 Water analogy model of capacitors and resistances.

It would be nearer to the electrical situation if the bucket were thought of as a pump that could suck water out of the right-hand side and add it to the left and if it could instantly reverse this process to transfer water from left to right. This is the situation that occurs with alternating or oscillating currents. Thus with high-frequency oscillating currents the individual pulses of current are so short that the vertical resistances, the vertical pipes in our analogy, have no effect on the current which behaves as if they were not there at all. Similarly, if the direction of flow is changing very rapidly there is insufficient time for the rubber diaphragm to be displaced very much at all so it cannot exert much resistance to current flow.

Using the analogy of the water-filled tanks separated by a rubber diaphragm for a capacitor has the further advantage that it serves to illustrate the factors that affect capacitance. Thus the capacitance would be increased by having a larger area of plates in the same way as larger water tanks would increase the capacity of our analogous model. Similarly, the quality of the dielectric, the relative permittivity, is equivalent to the quality or elasticity of the rubber and the thinner the dielectric the greater the capacity, in the same way that a thinner sheet of rubber is more readily distorted. Of course, if the pressure is too great it would rupture the sheet of rubber in the same way that a voltage beyond the specified breakdown voltage would force current across the dielectric.

Figure 4.17 is not simply an abstract example – it is the electric circuit that describes what happens when voltages are applied to the tissues through the skin. In fact it is ions and not electrons that move in the tissues, as will be seen later, but the principles are the same. The epidermis has a relatively high electrical resistance whereas the deeper tissues and the electrodes used to apply the current have much lower resistances. In this way the system acts like a capacitor with the epidermis as the dielectric. Some flow of charges occurs, mainly through the sweat glands. It can thus be modelled by a pair of capacitors with resistances in parallel. The circuit is completed through the low-resistance deeper tissues, as shown in Figure 4.17. This accounts for the fact that short pulses or high-frequency alternating pulses, which are simply a string of alternating short pulses, can pass more easily through the skin and into the deeper tissues than longer pulses or continuous direct current. In the latter case much of the electrical energy is used up in the skin, which has therapeutic implications (see page 69 in *Electrotherapy Explained*).

MAGNETIC EFFECTS

Consideration has been given so far to the electrical nature of charges but when electric charges move, and only when they move, a separate force is generated. This is the magnetic force. It acts in a direction perpendicular to the line of motion of the electric charges. The region in which this force is evident is called a magnetic field. Like the electric field between proton and electron it becomes weaker further from the magnetic source. Electric and magnetic forces are really two facets of the same phenomenon, that of electromagnetism, and together exhibit a third aspect – electromagnetic radiation.

Thus:

1. Electric fields are due to static electric charges.
2. Magnetic fields are due to moving electric charges.
3. Electromagnetic radiations are due to changes of velocity (acceleration) of electric charges (see Chapter 3, page 33 and Chapter 8, page 181).

Virtually all modern technology, the generation, transmission and application of electricity and radiations are based on electromagnetism.

Magnets

Permanent magnets (see page 66) are familiar as bar magnets, horseshoe magnets and, in a small light suspended form, as compass needles. The use of these latter is well known in that one end, the north-seeking end, points to the magnetic North Pole and the other to the South Pole of the earth. This occurs because the earth itself behaves as a gigantic magnet. The naming of the poles of the magnet is traditional because the end

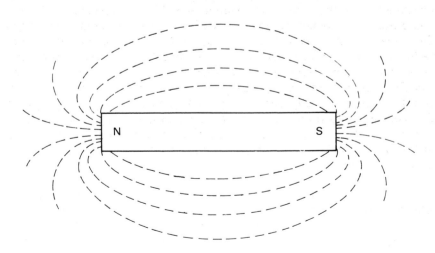

Fig. 4.19   The magnetic field of a bar magnet.

(a)

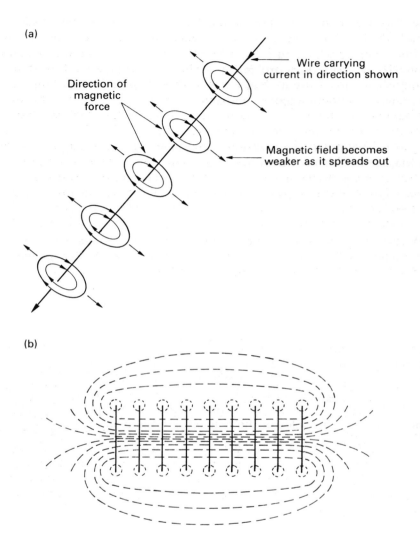

Direction of
magnetic
force

Wire carrying
current in direction shown

Magnetic field becomes
weaker as it spreads out

(b)

Fig. 4.20 The magnetic fields of an electric current: (a) around a straight wire carrying a current; (b) around a coil of wire carrying a current.

attracted to the north is called the north pole of the magnet and vice versa but, in fact, magnetically, north poles attract south poles and repel other north poles. This property of directionalism or polarity is important and a consequence of the direction of electron motion. Magnets are therefore said to be *dipolar*. The force of attraction or repulsion between the magnets is inversely proportional to the square of their distance apart. Plotting the direction of the magnetic force around a magnet by using a large number of little magnets, such as iron filings, gives a characteristic pattern, illustrated in section in Figure 4.19. This shows the greatest force close to the poles at each end. Activities that increase the random motion of molecules in the microstructure, such as heating or hammering, tend to destroy the magnetism of the steel bar due to disruption of the magnetic domains (see below). The lines shown in Figure 4.19 represent the magnitude and direction of the magnetic force serving, like the contour lines of a map, to describe steepness or strength of the magnetic field. They have, of course, no physical reality and only represent the magnetic field which is a vector quantity since it has both magnitude and direction. It will be noted that the shape of the magnetic field around a bar magnet is the same as the field around a long coil carrying a current, i.e. a solenoid, as illustrated in Figure 4.20b. An obvious comparison is that the magnetic field passes through air in the centre of the coil but through the metal of the bar magnet.

Measurement of magnetic effects

The force between two magnetic poles can be described in newtons and metres and called *magnetic flux*. It is measured in units called *webers*, and represented as the lines of magnetic force, as in Figure 4.19 and 4.20b. However, what matters is the force per unit area, called the *magnetic flux density* and measured in *webers per square metre*. One weber per square metre is called a *tesla* (or 10 000 gause). The flux density may be thought of as how closely the lines of magnetic force are crowded together. It can be seen in Figure 4.19 that they are spread apart beside the middle of the magnet but close together near the ends, indicating a stronger magnetic field at these places.

The electron as a magnet

Because each electron spins or rotates about its own axis it is, in effect, a loop of electric current and produces a tiny magnetic field. In other words, the electron acts as a miniature magnet. A clockwise-spinning electron has a magnetic field of one polarity and an anticlockwise one of the opposite polarity. In most atoms the electrons form spin pairs (see Chapter 2), which has the effect of cancelling out any magnetic effect. In certain atoms, however, there is an unpaired electron. An outside magnetic field can cause the tiny *domains* – separate small volumes of the material, typically only a few fractions of a millimetre in size – to line up all in the same direction, thus turning the whole piece of material into a magnet. When such material is not magnetized the domains lie in different directions and tend to cancel each other out. Ferromagnetic materials have these strong magnetic properties. They include the elements iron, nickel and cobalt, together with various alloys of these, notably steel, an alloy of iron and carbon. In soft iron, the alignment of domains can be achieved with very little energy; hence it can be easily magnetized, but will become demagnetized when the outside magnetic field is removed. If, however, it takes a good deal of energy to orientate the domains, as in steel, then it is difficult to demagnetize the material and a *permanent magnet* results.

ELECTROMAGNETISM

The other way of producing a strong magnetic field is to pass a current in a conductor. Such a device is called an electromagnet. When the electrons are in motion in a straight conductor, such as a straight piece of wire, the magnetic effect is exerted at right angles, spreading out in a cylindrical field around the wire (Fig. 4.20a). In order to illustrate the direction and strength of this magnetic field a large number of small magnets – iron filings or little compass needles – can be used. A larger and stronger magnetic field could be produced either by increasing the current or by putting more wires carrying current in the same region. In both cases the current per unit volume is increased. The usual way to form a strong electromagnet is to form a spiral coil of many turns close together (Fig. 4.20b). This concentrates the magnetic field by bunching together the lines of force. This can also be done by passing the field in a material such as soft iron. Thus even quite moderate currents passed in a tightly wound coil of many turns of wire around a soft iron core will produce a strong magnetic field.

If the direction of the current of an electromagnet is reversed, the polarity of its magnetism also reverses. Thus a piece of soft iron can have its magnetic polarity rapidly and repeatedly reversed by placing it in the rapidly reversing magnetic field produced by an alternating current.

Now if an electromagnet is put close to another magnet it could cause the other magnet to move, i.e. attract or repel it, if it were free to move. However as magnets can only exist as dipoles (with a north and a south pole), they can only rotate in a uniform field. If the field is not uniform there can be lateral movement as well. Thus an electric current can produce movement via a magnetic field. This is the principle on which electric motors work and much else besides. The reverse will also occur. If a magnetic field is moved in respect to a conductor it can cause an electromotive force to be set up in the conductor and if the charges in the conductor are free to move a current will flow. This is easy to illustrate by connecting a length of wire – a conductor with free electrons – to a suitable meter and moving a permanent bar magnet close to it. This effect occurs without contact between the magnet and the wire so it is called electromagnetic induction. The current is said to be induced in the conductor. Naturally this effect occurs whether the magnet is moved in respect to the wire or vice versa. Similarly, if the permanent magnet is replaced by an electromagnet which is then moved in respect to the wire, a current is again induced. Perhaps less obviously, it is found that current will be induced while the electromagnet remains stationary but the current through the electromagnet is switched on and off. This shows that it is movement of the magnetic field that causes the effect.

Experiments along these lines were first demonstrated in about 1831 by Michael Faraday, the great English physicist. When asked at one of his demonstrations what use it was, Faraday is reputed to have politely replied: 'Madam, of what use is a newborn baby?' He was surely right.

The growth and development of these concepts have led to the generation and distribution of electricity and a plethora of applications on a worldwide basis. Many of these applications involve the interaction between magnetic, electrical and mechanical energy and include such common devices as electric motors, loudspeakers, electric buzzers, microphones, dynamos, transformers and many other familiar devices.

Switches

A switch worked by an electromagnet (a relay) is perhaps the simplest example. Current energizes an electromagnet which attracts a piece of soft iron (an armature) that closes or opens a switch as it moves (Fig. 4.21a). These are often used to allow small currents to control circuits carrying much larger currents. Such devices are used with core-balanced coils (see Fig 4.36) as safety devices to cut the mains current if leakage is occurring, as described in *Electrotherapy Explained*, page 93. Similarly, reed switches are used in which two strips of metal conductor – reeds – are enclosed in a coil. Current through the coil causes opposite magnetic polarity in each strip so that they are attracted together to close the circuit (Fig. 4.21b). When a relay is made to open a switch in the circuit of the electromagnet itself, then the electromagnet is repeatedly switched on and off at a frequency which depends on the mechanical properties of the switch. The switch thus vibrates to and fro and can form an electric buzzer or bell, for instance (Fig. 4.21c). Suppose the current to the electromagnet can be made to vary in an appropriate way and the magnetic field produced is able to attract a thin sheet of metal (a diaphragm) – the motion of the metal can produce sound waves, as occurs in a simple loudspeaker or earphone.

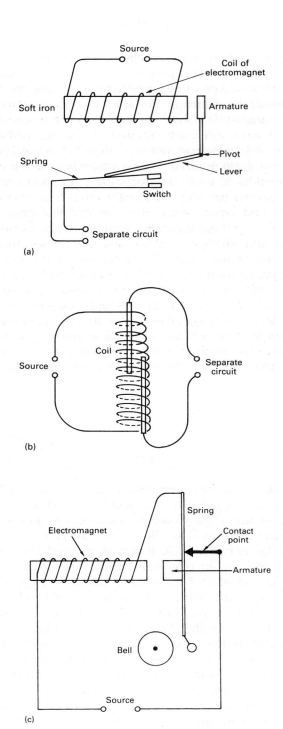

Fig. 4.21 Applications of electromagnetism: (a) a switch; (b) a reed switch; (c) an electric bell.

Meters

The force generated by an electromagnet depends on the current producing the magnetic effect so that a measure of the force will give a measure of the current. This mechanism is used to measure both current intensity and voltage. The most sensitive ammeters, milliammeters or voltmeters are of the moving coil type in which a small light coil is pivoted between the poles of a large strong permanent magnet. Rotational movement of the coil, which carries an indicator needle, is resisted by a spring (Fig. 4.22). The larger the current the stronger the magnetic field and hence the further the coil will turn against the resistance of the spring. The movement of the indicator needle is therefore a function of the current. A milliammeter can measure a few milliamps quite accurately and the same meter can measure larger currents by bypassing most of the current. Thus, if a low-resistance pathway is provided in parallel with the meter, the greater part of the current flows through the parallel resistance (see page 44), allowing a small current to pass through the meter. This small current is a fixed proportion of the total current, depending on the properties of the resistance of the meter and the parallel resistance. Thus if a meter which reads up to 1 mA on its scale were to be used to measure currents up to 1

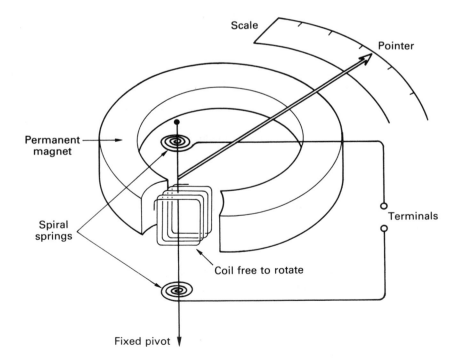

Fig. 4.22 A moving coil meter. N.B.: Most meters have a soft iron core. This is omitted for clarity.

A it would need a parallel resistance (shunt) so that 1 mA went through the meter and 999 mA went through the shunt. In other words, if the meter had a resistance of 100 Ω then a parallel resistance of approximately 0.1 Ω would be appropriate.

A milliammeter may also be adapted to function as a voltmeter, because the current is, of course, proportional to the voltage in any given circuit (see Ohm's law, page 41). In order to measure the voltage across any circuit it is necessary to have a high-resistance meter so that only a very small current can flow, which does not significantly alter the total current in the circuit. A large separate resistance is placed in series with the meter so that if, say, the total resistance including the meter is 1000 Ω, then 1 mA of current will flow and be measured by the meter when 1 V is applied. Similarly, if 2 V were applied then 2 mA would flow. Marking the scale in volts instead of milliamps turns it into a voltmeter.

In order to measure current in a circuit all the electrons must pass through the meter's coil or shunts. The meter must, therefore, be placed in series in the circuit. It must have a low resistance so that the current through the whole circuit is not reduced. A voltmeter, on the other

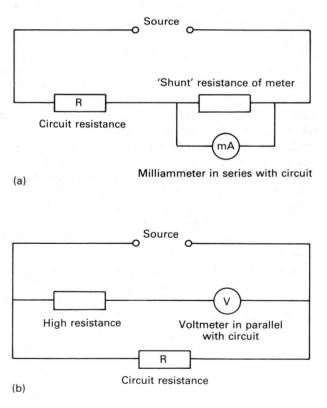

Fig. 4.23 Measuring current and voltage: (a) a milliammeter; (b) a voltmeter.

hand, must be placed in parallel with the circuit whose voltage it is to measure and have a high resistance so that only a small current is actually utilized for measurement. These ideas are shown in Figure 4.23.

If the current through such meters is reversed the movement of the coil will reverse also, so that the meter will determine the direction as well as the intensity of current. In many situations alternating current is to be measured and this requires either that the current be rectified (see later) and then measured with the moving coil meter or measured with a different type of meter in which only the strength and not the direction of the magnetic field is significant.

The electric motor

To extend the idea of a moving coil meter a little further leads to an electric motor. All that is added is a means of allowing continuous rotation. This is achieved by having a sliding contact connection made of small blocks of carbon, called brushes, pressing against copper segments called the commutator, which is in turn fixed to the rotating shaft and electrically connected to the coil. It is so arranged that as the coil and shaft turn, the commutator connection to the carbon blocks is reversed, so reversing the current in the coil at each half-turn. In this way the forces on the coil cause it to turn continuously. This principle is shown in a simple way in Figure 4.24. For practical electric motors many separate coils are used, wound on a soft iron core and each coupled to a particular pair of commutator segments. This arrangement leads to greater power and smoother rotation. Motors driven by alternating current work in a somewhat different way.

Electric motors of various sizes and powers are to be found everywhere in our technological society, large ones driving trains, moving lifts or escalators and working industrial machinery and smaller ones in the home turning extractor fans, hairdryer fans, working refrigerator or central heating pumps, food processors, electric drills and a host of other devices. In a car the engine is started by an electric motor as well as operating fans and windscreen wipers, for example. Somewhat smaller battery-operated motors drive the spools of tape recorders, desk fans and toys. Electric motors with some therapeutic or rehabilitation connection include those in powered wheelchairs, hoists, stairlifts and all the different pumps used for variable pressure devices, suction application of interferential therapy and in the hydrotherapy pool.

So far consideration has been given to the way electrical and magnetic forces produce mechanical energy but the reverse, that is, the generation of electrical energy from mechanical, also occurs. A device that converts mechanical to electrical energy electromagnetically is called a dynamo and is, at its simplest, an electric motor in reverse. Basically, if a conductor moves in relation to a magnetic field an electromotive force is induced in the conductor and if the circuit is completed, a current will

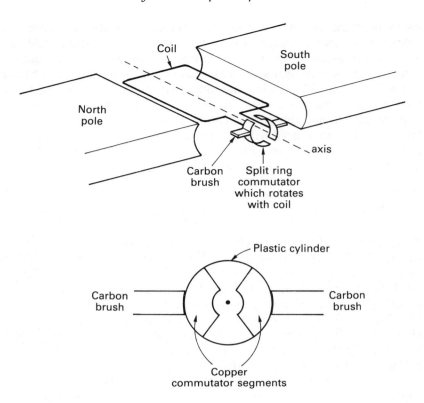

Fig. 4.24 The electric motor.

flow. The word *induction* is used to indicate that current is generated without physical contact between the parts. What happens is that the magnetic force acts on the free electrons of the conductor to cause them to move or tend to move. If the conductor, say a coil of wire connected to a milliammeter, is moved beside the north pole of a permanent bar magnet, a current is induced and the meter is seen to deflect. Moving the coil in the other direction causes the same brief burst of current while the movement is taking place but in the opposite direction. Similarly, reversing the direction of the magnetic field, e.g. moving beside the south pole of the magnet, reverses the current.

Using an electromagnet instead of a permanent magnet leads to the same effects except that the induced current can easily be made larger by putting a larger current through the electromagnet. It will also be found that, if the size of the conductor, the number of turns of the coil and, most importantly, the rate at which it moves are increased this will lead to larger induced currents. This concept is often expressed as *Faraday's law* and is essentially: the induced electromotive force is directly proportional to the rate at which the conductor cuts the magnetic field.

Less evidently, it will also be found that the greatest induced effects occur when the conductor and magnetic field move at right angles to one another.

As described, the direction of the induced electromotive force and hence of any current depends on the directions of the magnetic field and of the movement. It is easy to predict the direction of the current because it is always such as to oppose the changes producing it. This is usually described as *Lenz's law*, after the Russian scientist, Heinrich Lenz (1804–1865).

Fleming's laws

It may have been noted that the three forces being considered, mechanical, electrical and magnetic, act at right angles to one another in the electric motor or meter and in the dynamo. Although the directions of these forces can be worked out from first principles, they are conveniently summarized by using laws devised by Sir John Ambrose Fleming (1849–1945). The left-hand motor rule says that if the first finger, second finger and thumb of the left hand are stretched out at right angles to one another they represent the direction of the magnetic field, conventional current and motion respectively in an electric motor (*first* for *field*; *second* for *current*; thu*mb* for *motion*; Fig. 4.25). For a dynamo, converting mechanical to electrical energy, the meanings are similar but Fleming's right-hand rule is used.

Mutual induction

Returning to the production of an e.m.f. by moving a magnetic field relative to a conductor, it is worth considering what is meant by moving the magnetic field. So far the magnet, permanent or electromagnet, has been physically moved in relation to a conductor or vice versa. If the magnetic field results from the current in an electromagnet it can be varied by varying the current from no current, hence no magnetic field, to any strength of magnetic field, limited only by the current. This change of magnetic field is in effect movement of the magnetic force as far as the electrons of the conductor are concerned. Switching the current on and off through the electromagnet will cause a rising and falling magnetic field which acts on the electrons in a conductor, inducing an e.m.f. Thus if two coils of wire are placed close together and a varying current passed through one, it becomes an electromagnet with a varying field which influences the second coil, inducing an e.m.f. in it (Fig. 4.26). This is effectively the transmission of electrical energy between two stationary circuits by means of a moving magnetic field, a process called *mutual induction*. If the current through the first coil, the primary coil, is varied regularly – as occurs if an alternating current is used – then an equivalent alternating current will be induced in the other, secondary coil. This arrangement is the basis of the transformer, so called because it is able to change or transform voltages. Before

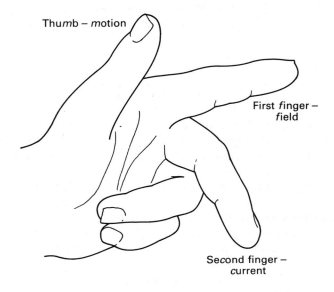

Thumb – motion

First finger – field

Second finger – current

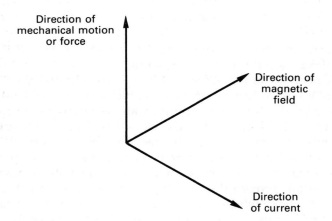

Direction of mechanical motion or force

Direction of magnetic field

Direction of current

Fig. 4.25 Fleming's left-hand rule for an electric motor.

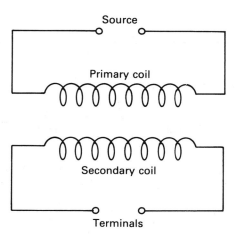

Fig. 4.26 Mutual induction.

discussing this important device it is necessary to consider the effects of varying magnetic fields in more depth.

Self-induction

If changes in current and thus magnetic field can affect the electrons of a nearby conductor, will these changes also affect the conductor carrying the current? After all, each section of the conductor will be affected by the magnetic field from an adjacent section. This process does occur and is known as *self-induction*. The consequence is that the current in the conductor is restricted in its rate of rise and fall. This is what would be expected from Lenz's law. Consider a simple coil of wire through which a current is passed by closing a switch (Fig. 4.27a). At the moment the switch closes, the current starts to rise and as it does so, the magnetic field it causes rises as well, cutting adjacent turns of the coil. The effect is to generate an e.m.f. that acts in the opposite direction to that driving the current; an effect opposing the change producing it, as expected from Lenz's law. Such an induced e.m.f. is often called a back e.m.f. This will interfere with the rate of rise of the current, causing it to rise exponentially, as shown in Figure 4.27b. The energy lost in the slow rise of current has gone into the magnetic field which now acts around the coil. If the current is now decreased the energy will be returned as a forward e.m.f. However, simply opening the switch in the circuit shown in Figure 4.27a would lead to a somewhat dramatic result because as the switch opens it causes a high resistance in the circuit and hence an abrupt fall of current in a very short time. This would result in a large forward e.m.f.; all the stored energy in the magnetic field would be returned more or less at once, which could produce a large spark across the switch contacts.

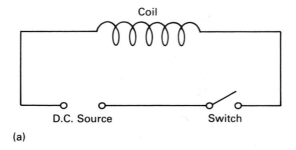

Coil

D.C. Source   Switch

(a)

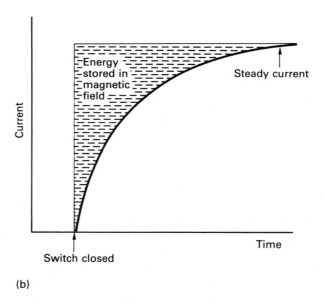

Current

Energy stored in magnetic field

Steady current

Switch closed

Time

(b)

Fig. 4.27 Self-induction: (a) a coil of wire through which a current is passed; (b) the exponential rate of rise of current.

The amount of self-induction that would occur for any given rate of change of current is determined by the size, shape and number of turns of the coil – a property known as *inductance*, often denoted by the letter L. The unit of inductance is the *henry*, named after the American electrical engineer Joseph Henry (1797–1878) who observed the effect of self-induction in 1832. One henry (H) is the inductance of a circuit such that a rate of change of current of 1 A s^{-1} produces an induced e.m.f. of 1 V. Inductance is thus a quality of the circuit, like capacitance and resistance. In any real circuit all these three factors operate to some extent and can be collectively called *impedance* (Z; see page 88). A device with negligible resistance and capacitance but with inductance is called an inductance. It is a coil of wire with or without a soft iron core (see later).

Eddy currents

The interaction between a moving magnetic field and a conductor does not only involve wires. The conductor could be a solid lump of metal or, as considered later, part of the human body. In a piece of metal, such as the iron core of an electromagnet, the induced e.m.f. can cause currents through enclosed pathways within the metal. These currents, because they often result from alternating magnetic fields and hence flow to and fro, are called *eddy currents* – they eddy to and fro. Such eddy currents can be quite large and cause heating and hence waste energy within the piece of metal. In many devices, such as electric motors, dynamos or transformers, energy losses are avoided by lamination of the iron core, i.e. it is built of a stack of iron plates each insulated from the next. As a result, the magnetic properties are only trivially diminished but current flowing across the plates is enormously reduced.

These induced electromagnetic effects, including the production of eddy currents, will only occur as a result of movement between a magnetic field and a conductor. If the field remains stationary with respect to the conductor then no effects will occur. This is not to say that a steady magnetic field will not influence a flow of electrons. Two important examples of this are the effect of a static magnetic field on a beam of free electrons, as occurs in a television set, oscilloscope or electron microscope (see page 111), and the *Hall effect*. This latter is the sideways force exerted on a flow of electrons by a static magnetic field. It is most evident in a block or slab of semiconductor material when a voltage develops across the material at right angles to both the magnetic field and the current. Since this voltage is proportional to the strength of the magnetic field a device consisting of a slab of semiconductor with a small constant current maintained across it and a sensitive voltmeter at right angles can be used to measure the strength of magnetic fields. This can be made quite small and is known as a *Hall probe*.

ALTERNATORS AND ALTERNATING CURRENT

The dynamo, considered above, would produce direct current but the familiar mains current is an alternating current (a.c.). Not only is this common to all countries but alternating current from sources other than the mains is widely used. It is produced by a dynamo converting mechanical to electrical energy but it is often, reasonably, called an alternator. Before considering how it works, it is sensible to explain why the apparently more complicated a.c. is universally used as the mains current.

Basically a.c. is used because it is so much more efficient. In the first place the alternator is simpler and cheaper to maintain than the equivalent d.c. dynamo. Secondly, as will be described below, the voltage can be easily changed or transformed by a transformer. This is an enormous advantage, enabling cheaper and more efficient transmission of electrical energy as well as providing electrical pressures or voltages appropriate to the task in hand. Finally it avoids the problems associated with the chemical changes produced by d.c. and described on page 98. Switches and contact points carrying large direct currents rapidly become encrusted with deposits caused by electrolysis which reduce their efficiency. Evenly alternating currents are obviously free of this problem, allowing smaller and cheaper contacts and connections to be used.

Generation of alternating currents

A simple form of alternator is a coil of wire rotating in a magnetic field, as shown in Figure 4.28a. It is the same as the electric motor, except that the commutator is replaced by a pair of *slip rings*, i.e. little cylinders. Each end of the coil is connected to a slip ring which rotates in contact with a block of hard carbon, a carbon brush. The slip rings are fixed to the axle or shaft, but insulated from it, so that they rotate with the coil and can lead current to and from the coil. In the generator the coil is being moved in the magnetic field by some external force. If point A of the coil (Fig. 4.28a) is being moved out of the page and the field is from north (N) to south (S), by Fleming's right-hand rule, the current flows clockwise round the coil. If the small segment of coil, depicted in Figure 4.28b, is imagined to be rotating as shown, it will move rapidly in respect to the magnetic field as it goes up and down and will cut through many lines of force but will not cross the magnetic field as it moves parallel to it at the top and bottom. The induced e.m.f. will therefore be large where the conductor moves vertically at right angles to the magnetic field but zero at the top and bottom where the conductor is moving horizontally, parallel to the field. Between these points the induced e.m.f. will vary (as the sine of the angle of rotation). The resulting induced e.m.f., and hence current if the circuit is complete, will follow a sine wave, as indicated in Figure 4.28c. By considering the discussion in Chapter 3 on the relationship between wave and cyclical motion, it can be seen that rotating a coil of wire in a magnetic field will lead to the production of an evenly alternating e.m.f. following a sine waveform repeated at the frequency of rotation.

In practical a.c. alternators for the mains current the situation is reversed. The coil, called the *stator*, is fixed and the electromagnet rotates and hence is called the *rotor* (Fig. 4.29a). This has the advantage that the slip rings and brushes only have to carry the relatively small current for the electromagnet while the a.c. output is taken from the stator. It is obviously more economical if one alternator can supply three separate circuits simultaneously. This arrangement, illustrated in Figure 4.29b, is used to generate a three-phase mains current. The stator coils, shown opposite one another, are continuous so that the electromagnet

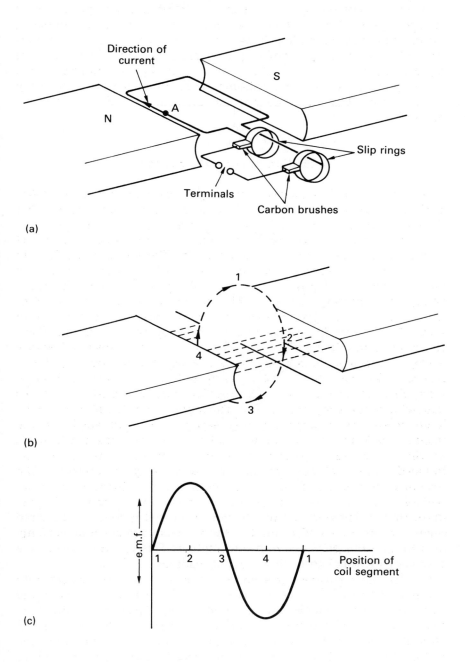

Fig. 4.28 A simple alternator – a coil of wire rotating in a magnetic field

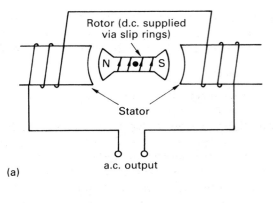

(a)

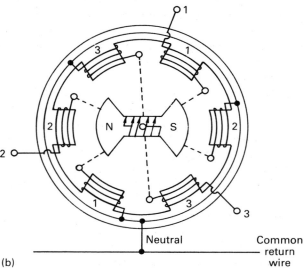

(b)

Fig. 4.29 (a) A rotor; (b) a three-phase alternator showing three stator coils.

(rotor) moves between the two halves of the coil in each of the three coils. This arrangement means that the three circuits generate currents that are 120° out of phase with one another. This not only allows the alternator to work more smoothly but leads to economies in distribution. The three currents are out of phase and each has a separate distribution lead. However, as, at any one time, the sum of the three currents is zero, a single return wire carrying a very small current is sufficient. Large electrical users, such as industrial installations, take all three phases from the supply to operate special three-phase electric motors and other devices. Such motors are simply the alternators of Figure 4.29 in reverse and are synchronous with the 50 Hz frequency of the mains and thus rotate at a constant speed. Domestic consumers are supplied with a single phase but the number of houses supplied is roughly balanced for each phase in any district. Further consideration is given to the domestic distribution of the mains on page 89.

Measurement of alternating current

The regular variations of alternating current along the horizontal time scale allow the frequency of current or voltage to be described as so many complete cycles per second or hertz. The mains in many countries, including the UK, is about 50 Hz but in others, such as the USA, it is 60 Hz. Since both the e.m.f. and consequent current are constantly varying it is not easy to assign a single value to describe either. The peak value, peak voltage or peak current is used sometimes but this only occurs twice in each cycle. An average value would be of more use but, of course, a simple average would be useless for an evenly alternating current since it would always be zero. What is used in practice is the effective or *root mean square* (rms) value. The effective (rms) value of an alternating current is the same as the intensity of direct current that gives the same power. Figure 4.30 shows parallel graphs of voltage, current and power for an alternating current passing in a circuit which has only resistance. Because there are no capacitive or inductive effects, the voltage and current are in phase with each other. As power is dissipated in the resistance, say as heat, the rate of heat production – the power – varies with the changes in voltage and current.

Power measured in watts = e.m.f. measured in volts × current measured in amperes

The direction of the e.m.f. and current makes no difference to the power, so output is described by the unidirectional waveform. The average power is the mean of these waveforms. For sine curves it can be shown mathematically that the effective value of the current is equal to the peak value of the current divided by the square root of 2, i.e.

$$\text{Effective current} = \frac{\text{peak value}}{\sqrt{2}}$$

and similarly for the voltage:

$$\text{Effective voltage} = \frac{\text{peak value}}{\sqrt{2}}$$

$$(\sqrt{2} = 1.414, \frac{1}{\sqrt{2}} = 0.707)$$

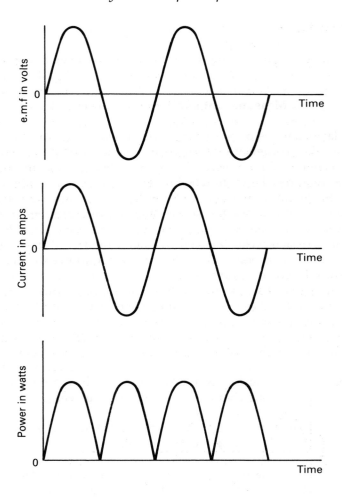

Fig. 4.30 Parallel graphs of voltage, current and power for an alternating current passing in a circuit which has only resistance.

The effective or rms current, or voltage, is thus 0.707 of the peak current or voltage and the peak current or voltage is 1.414 times the rms current or voltage. As mentioned, the reason for using the rms value is that it gives the same power as a d.c. source of the same value. The mains in the UK are provided at 230–240 rms voltage so that at 240 V rms the peak voltage is approximately 340 V (240 V × 1.414 = 340 approximately). Therefore mains current in the UK and many other countries is a 50 Hz evenly alternating current of sinusoidal waveform with an rms voltage of around 240 V and peak voltages of about 100 V more, supplied domestically as one of three phases generated for distribution.

TRANSFORMERS

If sinusoidally varying voltages, as described above, are applied to a circuit the current will follow the same variation (Fig. 4.30) and the resulting magnetic field will change in the same way. Such a continuously varying magnetic field will cause an induced e.m.f. in any nearby conductor. The induced effect will be greatest when the magnetic field is changing rapidly and least when it is stationary. Consideration of Figure 4.30 shows that the current, and hence the magnetic field, changes rapidly as the current changes direction – as the graph of current crosses the zero line – but is stationary for an instant at each peak of the graph as the current stops rising and starts to fall. The resulting induced e.m.f. will therefore be at maximum during the rapid change but zero when no change of magnetic field occurs. Between these points it will alternate sinusoidally but out of step with the primary current, as shown in Figure 4.31. (The graph of induced e.m.f. is a plot of the rate of change of the current.)

An arrangement typical of a low-frequency transformer appropriate for mains current is shown in Figure 4.32 in which two coils of wire are wound over one another around a central laminated soft iron core. Of course, real transformers consist of many hundreds or thousands of turns rather than the six shown here for clarity. The iron core serves to

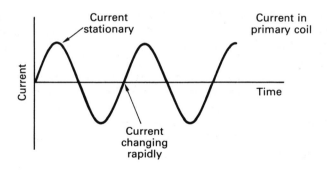

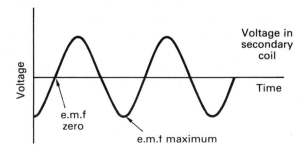

Fig. 4.31 The relationship between primary current and secondary voltage.

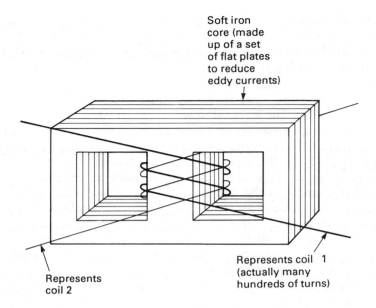

Soft iron
core (made
up of a set
of flat plates
to reduce
eddy currents)

Represents coil 1
(actually many
hundreds of turns)

Represents
coil 2

Fig. 4.32 A low-frequency transformer.

increase the magnetic field and the arrangement shown in Figure 4.32 decreases losses by reducing leakage of the magnetic field by retaining it in a complete magnetic circle. The core is laminated, i.e. built up of separate flat iron plates with insulation between them in order to diminish the eddy currents that are inevitably produced in the iron core (see page 77).

The two coils, labelled 1 and 2 in Figure 4.32, are termed primary and secondary, referring to input and output. The changing magnetic field about the primary coil affects all turns of both primary and secondary equally. This produces a back e.m.f. in the primary coil due to self-induction (see page 75) and an e.m.f. in the secondary whose magnitude depends on the number of turns affected. Thus if there are twice as many turns of secondary coil as there are in the primary the induced e.m.f. will be twice that of the primary. In this case it would be called a step-up transformer. Fewer turns in the secondary would give a lower induced voltage and it would be a step-down transformer. The same piece of apparatus can act in either way depending on which coil takes the input current. A step-down transformer, reducing the mains voltage from 240 to 12 V to drive a few tens of milliamps for something like a therapeutic stimulator, would have a 20:1 ratio of turns.

The output power cannot exceed the input power so the product of voltage and current in the secondary will be the same as that in the primary coil. (Actually it will be a little less because there will be losses, which are considered later.) Thus watts in primary coil (input) = watts in secondary coil (output). This gives rise to the expression 'watts in = watts out'. For example, consider a perfect transformer with 100 turns of primary coil through which the 240 V mains is driving a 4 A current. This is a power input of 960 W (240 × 4). A secondary coil, having 400 turns, would have a voltage of four times the primary, i.e. 960 V. Thus the current would be:

$$\frac{\text{watts}}{\text{volts}} = \frac{960}{960} = 1 \text{ A}$$

An even simpler way of considering the relationship is to say that if the voltage is increased fourfold the current is quartered.

Clearly the transformer is altering the relationship between pressure and flow of electric charges. It is not a case of something for nothing; the total energy remains the same. In the case above a large flow (4 A) at low pressure (240 V) is changed to a small flow (1 A) at higher pressure (960 V).

It has been noted already that self-induction occurs in the primary and it will be evident that flows of current in both primary and secondary coils will generate magnetic fields that will interact. This interaction between the two coils is referred to as *mutual induction*. The actual current in the secondary circuit depends on circuit resistance (Ohm's law) and this will dictate the current through the primary coil. Now, if there is infinite resistance in the secondary circuit, e.g. an open switch in the circuit, then no current would flow and energy dissipation would be nil. There would be zero power. In these circumstances there would only be a very small primary current. If the resistance of the secondary circuit is reduced, current flows and the magnetic conditions of the transformer so adjust themselves as to allow sufficient primary current to provide an appropriate amount of power.

There is always some energy loss between the input and output of all transformers, as already mentioned. These arise from three sources:

1. The ohmic resistance of the coils.
2. The small eddy currents that circulate within the laminations.
   Both of these lead to heating depending on the current intensity, in accordance with Joule's law. Thus, heating is proportional to I^2Rt, where I is current, R is resistance and t is time. If I is measured in amps, R in ohms and t in seconds, the amount of heat produced will be in joules.
3. The heating due to repeated reversal of magnetism in the core. This is due to the internal friction generated by the repeated rotation of the molecular magnets of the iron (see page 65). Its magnitude depends on the type of iron and is called a hysteresis cycle.

These losses are minimized by using appropriately thick low-resistance copper wire for the coils that will take high currents and, as noted earlier, reducing eddy currents by limiting the thickness of low-resistance iron in which they flow. This is effectively reducing current by placing a high resistance – the insulation between the iron plates – in its path. The hysteresis losses can be limited by the use of an appropriate alloy of iron. Good design can thus make transformers quite efficient. This is especially so for large transformers which can be 95–99% efficient. The lost energy appears as heat (and some in the form of mechanical vibration as some transformers are heard to buzz). Small transformers become slightly warm during operation. Large ones, even the most efficient, need special cooling arrangements, such as circulating cooling liquids or immersing the coils in tanks of oil. (Even a 1% loss in a large transformer is a lot of heat – 1% of 1000 kW is 10 kW, equivalent to 10 single-bar electric fires.)

Practical low-frequency transformers are often constructed with several secondary windings giving different voltages for different circuits in one piece of equipment. These are effectively several separate transformers combined into one piece of apparatus.

Due to their economic importance and widespread use, low-frequency transformers are widely recognized and discussion of a transformer is often taken to mean a low-frequency apparatus. It must be understood that the same principles can be applied to any regularly varying current so that high-frequency currents can be easily transformed from one circuit to another by means of a pair of conductors placed close to one another. This idea is utilized in radio receivers and shortwave diathermy apparatus, as discussed in Chapter 5. It has already been stressed that induction depends on the rate of change of the magnetic field and hence of the current. High-frequency current will have much higher rates of change and will therefore generate much more self-induction (see also page 75). This means that high-frequency transformers are built to have much lower inductance and consist of a pair of coils with, perhaps, only four or five turns each for megahertz frequencies. Naturally no soft iron core is needed. The relationship of frequency with inductance is discussed on page 88.

Much of what has been described about electromagnetic induction, especially the transformer, is the result of an early development – the induction coil. This device was the earlier form of the transformer, being a way of producing high voltages and a means of progress towards the discovery of radio, X-rays and the electron. Basically, the induction coil is a step-up transformer with a primary and much longer secondary coil both wound on a soft iron rod. There is a means to open and close the primary circuit with a contact breaker on the same principle as the electric bell or buzzer, as already considered and shown in Figure 4.21c (see also Fig. 4.33). As the primary circuit is broken the magnetic field around the primary collapses rapidly, so inducing a high e.m.f. in the secondary.

Although Michael Faraday in the UK and Joseph Henry in the USA originated the production of electrical energy from magnetism and motion, it was an Irishman, N. J. Callan, a priest turned physics teacher

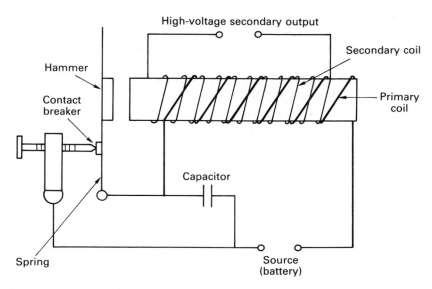

Fig. 4.33 An induction coil.

of County Kildare, who devised the basic form of the induction coil in 1836. This was followed by more, larger coils built by Callan and others during the 1840s and 1850s incorporating improvements and modifications, notably the provision of a capacitor across the contact points to diminish sparking (Fig. 4.33) and various ways of altering the secondary voltage smoothly by moving one coil over the other or inserting and removing the iron core. Interestingly, this form of the induction coil was in widespread use for over a century with essentially no further modifications as a source of therapeutic faradic current. The high spike of voltage can be raised to about 40 V and made to last rather less than a millisecond. The contact breaker spring oscillates around 60 Hz, which gives a set of pulses quite effective for stimulating motor nerves through the skin. (See *Electrotherapy Explained*, page 33 and Figure 3.4c.) The other use which has survived almost to the present is in the induction coil of the petrol car engine to provide the ignition spark.

In the latter half of the 19th century, larger and larger induction coils were built to give higher secondary voltages. The largest was made by A. Apps, a London instrument maker, in 1876 for William Spottiswode who used it to study gas discharges. It had an iron core 1.1 m long and a secondary coil of about 330 000 turns. The sparks produced between open electrodes could be over 1 m long (Shiers, 1971)! Induction coils were, for many years, the main source of high voltages for laboratory purposes. Röntgen discovered X-rays using one and Hertz first demonstrated radio waves being generated by one. 'Cathode rays' were shown to be negative charges (electrons) by J. J. Thomson, also using such a coil.

IMPEDANCE

It has been noted when considering self-induction that a varying current will induce a back e.m.f., by Lenz's law, in any conductor. The strength of this opposition will depend on the inductance (L) of the conductor and the rate of change (f) of the current. The faster the rate of change the greater the reaction. This is called *reactance* (X). The effect will also be greater with larger coils. In fact, if the coil is large enough and the frequency high enough there will be complete obstruction to the passage of the current.

It will be recalled that a capacitor behaves in precisely the opposite way. In other words, reactance is less with increasing frequency of current oscillation.

The term reactance refers to the difficulty of flow of an alternating current due to the inductance and capacitance of the circuit. The circuit will also have resistance due to the nature of the conductor and the temperature – ohmic resistance, as discussed on page 41. Thus for an alternating current there are three sources of resistance, all measured individually or collectively in ohms. They are together called *impedance* and given the symbol Z. Thus the impedance of a circuit to any given alternating current would depend on:

resistance, R – nature and temperature of conductor
reactance, X – inductive
           – capacitive

$$Z^2 = R^2 + X^2$$

For an alternating current inductance increases directly with frequency and conversely a direct current having zero frequency has no inductive reactance. A capacitor, on the other hand, has infinite impedance for a direct current but this diminishes with rising frequency. Thus:

$$Z \propto f \text{ for an inductance}$$

$$Z \propto \frac{1}{f} \text{ for a capacitor}$$

The impedance in ohms for a given inductance can be found from:

$$Z = 2\pi f L$$

where L is the inductance in henries.

Similarly, the impedance of a capacitor can be found from:

$$Z = \frac{1}{2\pi fC}$$

where C is the capacitance in farads.

The above has important practical consequences. For instance, inductive impedance can be used to block high-frequency currents while allowing low-frequency or direct current to pass easily, thus acting as a filter. When used in this way a simple coil of wire of appropriate dimensions is used called a *choke coil*. Where high-frequency currents are to be blocked a small coil will have enough inductance. To block lower frequencies a larger coil, with a soft iron core to increase the inductance, will be needed. Thus high-frequency and low-frequency chokes are of different construction. The same principle can be used to regulate alternating currents. A suitably-sized variable choke acts like a variable resistor (see page 46), but has the advantage that there is little loss of energy being converted to heat as happens with simple resistances. If a current path has significant capacitance, such as the path between cutaneous electrodes and the tissues (see page 62 and Fig. 4.17) then the impedance will decrease with higher-frequency currents. This is described on page 57 and explains why interferential current of 4 or 5 kHz will pass more readily through the skin than low-frequency or direct currents.

MAINS CURRENT

As noted already, the mains current for domestic use in the UK is an evenly alternating sinusoidal current of 50 Hz at 220–240 rms voltage. (In the USA 60 Hz at 110 V is provided.) Dynamos at power stations convert mechanical into electrical energy. The mechanical energy can be derived directly from falling water, i.e. hydroelectric power, or from a wind turbine. Most generators, however, are steam turbines driven with the high-pressure steam from burning coal, oil, gas or from nuclear fission. Such power stations can be quite efficient, with about 35% of the energy available in the coal or other fuel being converted to electric power. Even higher efficiencies can be achieved if some of the waste heat is utilized by industry or for domestic heating.

As described earlier, a three-phase system is used and stepped up to high voltages of several hundred thousand volts for transmission across country. The high voltages allow relatively low currents to pass, hence there is very little loss due to heating the transmission lines. A further economy is to use a single line at a high potential alternately positive and negative to earth. Cables suspended on pylons are a familiar sight in rural areas. In towns the cables are usually buried underground. At transformer substations the voltage is reduced to around 11 kV or 6 kV for large industrial uses or to the familiar 240 V supply for domestic users.

Alternating current is supplied to the building by a live wire and a neutral which is earthed at the transformer (Fig. 4.34). A meter is provided to measure the energy used in kilowatt hours or joules (1 kilowatt hour = 1000 W for 1 hour = 3.6 MJ). Main switch and separate circuits for lighting, which carry quite small currents, and power, which carry larger currents, are provided. In general lighting circuits have 5 A fuses while power circuits are protected by 13 A fuses. These latter are

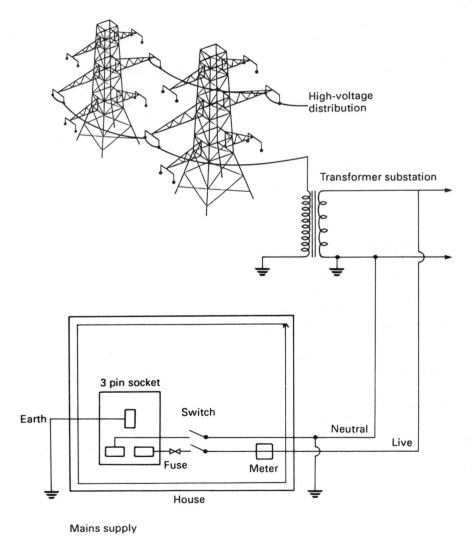

Fig. 4.34 Mains supply.

connected by three-pin plugs and are used for apparatus taking higher currents and with metal casing, including much physiotherapeutic apparatus. In many sites such circuits are connected to a ring main which doubles the capacity of the cable and is used with fused plugs. *Fuses* are effectively a weak point in the circuit, provided so that any current above a predetermined value will cause the short piece of wire to become hot enough to melt and thus break the circuit. Most modern apparatus that contains fuses utilizes the convenient cartridge type with the wire enclosed in a small tube. To understand the operation of fuses it is necessary to realize that they are in series with the circuit they are protecting so that all the current in that circuit must flow in the fuse wire. If the resistance of the rest of the circuit falls then the current rises (Ohm's law) and the fuse wire becomes hotter, which raises its resistance causing still more heating, leading to rapid melting.

The three-pin plug and socket

These are designed to connect only in the proper orientation so that the live, neutral and earth wires are always correctly connected. The earth pin is the longest so that it inserts first and is removed last. This ensures that there is no time when the live is connected but the earth is interrupted. The wiring is identified by the colour of the insulation: the live is brown, the neutral blue and the earth wire is yellow and green. These colours have the advantage of being easily distinguished by colour-blind people.

ELECTRICAL SAFETY WITH MAINS CURRENT

All electrical equipment working from the mains is made to be safe in one of two ways. For small portable pieces of equipment, such as radios or reading lamps, double insulation is used. The casing is made of some non-conducting, plastic material and the electrical conducting parts are separately insulated – *double insulation*. For apparatus with metal casing or to carry larger currents the fused three-pin plug with earth connection is used. The metal casing is connected to the earth wire of the three-pin plug and socket (Fig. 4.35) so that if the live wire were to contact the metal casing of the machine a large current would flow to earth, causing the fuse to 'blow' and thus disconnect the circuit. In the absence of an earth, or if it were broken or not properly connected, the metal casing of the machine could become live so that anyone touching it would provide a low-resistance path to earth and hence receive an electric shock. For this to happen two faults must occur at the same time, which is unusual. However it must be recognized that a break in the earth wire is not

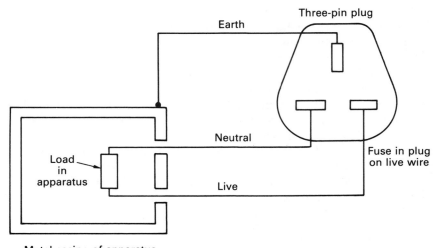

Fig. 4.35   Connections of three-pin plug to apparatus.

evident – it carries no current during normal operation – so this fault can exist unrecognized for quite some time. It is therefore important that the wiring and connections of apparatus are checked regularly.

It will be noticed that if only low currents are passing through the casing to earth, via someone touching it for example, the fuse does not necessarily blow because the total current is within its limit. In order to prevent such small leakage currents a core balanced relay may be incorporated into the circuit. Normally current in the live and neutral wires will be equal but if some current is leaking to earth the current in the live will be greater than that in the neutral. The two wires, live and neutral, pass through a magnetic core around which is wound a sensing coil (Fig. 4.36). The current in each wire will be equal but in opposite

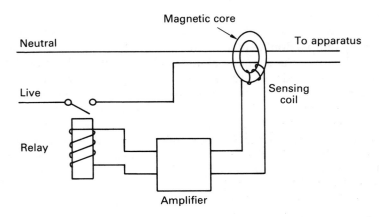

Fig. 4.36   A core balance relay.

directions so that no induced effect occurs, but if a smaller current passes in the neutral the difference induces a current in the sensor coil which, after being amplified, operates a relay switch to open the circuit (a further device that relies on electromagnetism).

From what has been described and from Figure 4.34 it is evident that current can pass from the live to earth through a person, which constitutes a danger. This can be much reduced by using an isolating transformer, i.e. a transformer in which the output is not earthed. A person connected to earth touching one terminal of the secondary coil of the transformer would not now receive a shock. Since the circuits of many pieces of therapeutic apparatus are from the secondary of a transformer this renders the current 'earthfree' and hence safer. To make this arrangement still safer a device to detect any earth leakage from the transformer can be incorporated.

PRODUCTION OF AN E.M.F.

An e.m.f. able to drive current through a circuit may be achieved in several different ways. So far only the direct conversion of mechanical energy to electrical energy, by electromagnetic induction, has been considered. However, other forms of energy can be converted directly to electrical energy.

The thermoelectric effect

If a circuit is formed with two different metals and the two junctions are at different temperatures, then a small e.m.f. is generated proportional to the temperature difference. Such a pair of junctions are called a *thermocouple* and are often made of the metals antimony and bismuth to give a good voltage, although constantan or iron and copper will also work well. A single junction will generate only a tiny e.m.f. – a few millivolts for a difference of 100°C – so that many junctions are usually coupled together to give a more substantial voltage. Due to the relatively trivial currents generated by quite large temperature differences these devices have no place in practical electrical generation but are used as temperature-measuring devices (see Chapter 7).

The photoelectric effect

Radiations, such as ultraviolet or visible radiations, can cause the emission of electrons from certain metals (discussed in Chapter 8) or the movement of electrons in certain semiconductors (see page 106). These effects have been widely used for measurement of radiations (light-meters) and in control systems such as automatic switches and burglar alarms as well as in television cameras as the basis of devices called photocells. Large arrays of such devices are used to provide power for some satellites in space and are in use in some parts of the world for the direct generation of electricity from sunlight.

The electric cell

Devices that convert chemical energy to electrical energy are called electric cells and when several are connected together this is known as a *battery* of cells. (In the vernacular it has become customary to refer to both single and multiple cells as a battery.)

If two different metal sheets are dipped into a bath of electrolyte (a solution containing ions that conducts a current and which may be an acid, base or salt) and a connection made between the two metals a current is found to flow between them. Chemical changes occur as the metals and electrolyte interact and it is energy derived from these that sets up an e.m.f. between the two metals to drive a current. For example, if copper and zinc electrodes are placed in a solution of dilute sulphuric acid a wire connecting them outside the solution would allow a flow of electrons from zinc to copper (described as current flowing from copper to zinc). In practice such a cell would quickly cease to function effectively due to the deposition of hydrogen bubbles on the copper plate, a process called *polarization*. With some modification (known as a Daniell cell) it can provide an e.m.f. of 1.1 V.

Other metals or metalloids and other electrolytes can be used, some providing rather higher voltages. The most widely used are zinc and carbon with ammonium chloride as the electrolyte, in the form of a paste. This is known as a sealed dry Leclanché cell (after Georges Leclanché, 1839–1882). The outer zinc can forms the negative electrode and the central carbon rod the positive (Fig. 4.37). The rod is surrounded by a mixture of manganese (IV) oxide and powdered carbon. The manganese (IV) oxide oxidizes the hydrogen formed, acting as a depolarizer. The jelly or paste of ammonium chloride is prevented from drying up because it is sealed in. These cells will deliver an e.m.f. of 1.5 V and are very widely used in torches, calculators, portable radios and cassette players and for transcutaneous electrical nerve stimulators (TENS). Depending on usage and conditions they can last for weeks, months or even years but eventually they become exhausted. Even without use they slowly deteriorate due to unavoidable local reactions in the cell.

Mercury cells, involving steel and zinc electrodes in a potassium hydroxide electrolyte, deliver a slightly lower e.m.f. of 1.4 V but last about twice as long as an equivalent-sized Leclanché cell. They also maintain a more nearly constant voltage throughout their life. They are used in hearing aids, watches and other small electronic devices and are often known as long-life batteries.

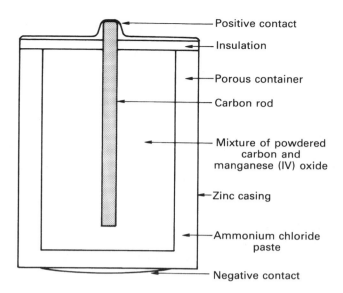

Fig. 4.37 A dry sealed Leclanché cell.

The Weston cadmium cell is designed to deliver an exact unvarying e.m.f. over many years and is used to calibrate laboratory equipment.

Fuel cells are larger and offer a more sophisticated method of producing electricity directly from chemical action. The best known involve gaseous hydrogen and oxygen combined over catalytic electrodes to provide electrical energy and form water. The fuel is continuously replenished.

All the foregoing cells are usually classed as primary cells to distinguish them from the secondary or storage cells described below.

Storage cells

In these cells the chemical action is reversible so that electrical energy may be used to effect a chemical change which remains stored, more or less indefinitely, until the electrodes are connected to a circuit allowing a flow of current as the chemical changes reverse. Thus the cell can be repeatedly charged and used, acting as a device for the conversion of electrical energy to potential chemical energy for storage and conversion back to electrical energy when needed.

The best known is the lead-acid storage cell familiar as the 12 V car battery which is, in fact, six 2 V cells connected in series. These cells have grid-like electrodes, to give a large surface area, made of a lead-antimony alloy. The positive plate is covered in lead (IV) oxide and the electrolyte is dilute sulphuric acid. As current is drawn from the cell both plates or electrodes become coated with lead sulphate and the electrolyte becomes more dilute. On recharging, one plate returns to lead oxide and the other to lead. Such cells are, rather curiously, called *accumulators*.

Nickel-iron (Nife) cells are similar. The negative plate is iron and the positive nickel oxide, with the electrolyte being a solution of potassium hydroxide. These cells have a lower voltage, about 1.2 V, and are more expensive than lead-acid cells but they are much more robust and may be left in a discharged state for long periods without deterioration. They are widely used in standby situations to provide emergency lighting in public buildings and hospitals. Nickel-cadmium (Nicad) cells are more expensive but more efficient and are often found as rechargeable batteries for small items such as electric razors or for some TENS sources.

There are many other cells under development, largely stimulated by the economic and environmental advantages of electric cars and rail transport. Zinc-air accumulators involve the conversion of zinc to zinc oxide using oxygen from the air through a porous nickel electrode. The electrolyte is potassium hydroxide. Other types involve sodium and sulphur or lithium and chlorine but need to operate at high temperatures of several hundred degrees Celsius. All these types would have very much higher energy densities than lead-acid cells but are very much more complex and expensive.

The cells in common use are summarized in Table 4.4. The storage of electrical energy on a small scale is effected by the storage cells noted above but on a large scale it is impossible. Mains electricity is used as it is produced, hence much expensive plant remains idle when the need for electricity is low.

Table 4.4 Different types of cells

Name and type	Electrodes		Electrolyte	Voltage	Uses
	+	−			
Primary cells					
Dry, sealed Leclanché	Carbon	Zinc	Ammonium chloride	1.5	Calculators Torches
Mercury cell	Steel	Zinc	Potassium hydroxide	1.4	Long-life hearing aids
Secondary or storage cells					
Lead-acid accumulator	Lead (IV) oxide	Lead	Sulphuric acid	2	Car battery
Nife cell	Nickel oxide	Iron	Potassium hydroxide	1.2	Emergency lights
Nicad cell	Nickel oxide	Cadmium	Potassium hydroxide	1.2	Rechargeable batteries

Biological tissues, notably nerve and muscle, are also able to generate electrical from chemical energy. This is principally used for signalling as intracellular communication and, of course, nerve impulses which form the basis of all neuronal activity. Some kinds of fish have developed special cells to generate a strong brief electric pulse which can be used to stun their prey. They are specially modified muscle cells which are relatively short and flat, called electroplaques, each of which can generate an action potential of some 0.1 V. Stacks of several thousand such cells can generate a momentary pulse of a few hundred volts. The electric eel can generate a pulse of about 300 V which can be used to stun prey and discourage predators.

Figure 4.38 summarizes energy conversions.

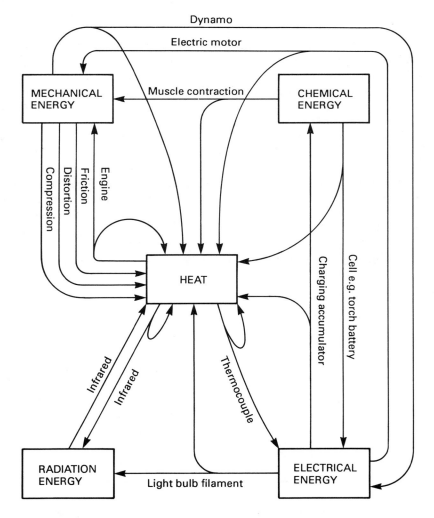

Fig. 4.38 Energy conversions.

CURRENTS IN LIQUIDS

The way in which current passes in liquids is very different from the mechanism described earlier for solid conductors. It involves the bulk movement of particles rather than electrons flitting from atom to atom and is therefore called a *convection current*. It is, of course, the way in which currents travel in the body tissues.

Many compounds break up into pairs of oppositely charged ions when they go into solution. This makes the solution into a conducting medium which is called an *electrolytic solution* or an *electrolyte*. Many crystalline compounds are made up of ions which split apart or *dissociate* when the solid is dissolved in water. Inorganic compounds, salts, acids and bases form electrolytic solutions but some substances – sugar and alcohol, for example – are non-ionic and dissolve without splitting, forming non-electrolytic solutions.

In electrolytes an electric current can occur due to the motion of positively and negatively charged ions in opposite directions. It will be recalled that ions are atoms or groups of atoms that have either lost an electron or two, rendering them positively charged, or gained one or two electrons, giving them a negative charge. Thus a solution of common salt containing positively charged sodium ions (Na^+) and negatively charged chlorine ions (Cl^-) would pass a current when an electric field was applied by the simultaneous motion of Na^+ to the negative region of the field and Cl^- to the positive region. In Figure 4.39 the electric field is provided by two metal plates in the electrolyte called electrodes. This is the arrangement commonly described when considering convection currents in liquids but it must be realized that the same thing will occur if the electric field is provided by a voltage applied to plates outside the solution, as occurs with shortwave diathermy.

To apply the electric field in Figure 4.39, metal plates or wires are used

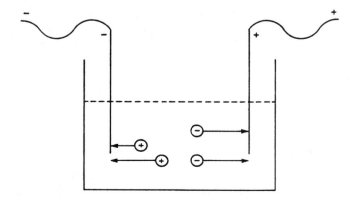

Fig. 4.39   The movement of ions when an electric field is applied to an electrolyte.

in contact with the solution. When a direct current is considered, the positively charged electrode is called the *anode* and the negative the *cathode*. Thus positive ions move towards the cathode and hence are sometimes, rather confusingly, called cations. Similarly negative ions may be called anions because they migrate towards the anode. In this text these terms will be avoided.

So that a continuous current may flow, the ions must be removed and replaced as they reach the electrodes. This can happen in the following ways. The ions are discharged, that is to say, they gain an electron from or lose one to the electrode, which occurs when the electrodes are made of chemically inert material like platinum, or new ions are formed from the material of the electrodes which then enter the electrolytic solution. In this latter case the electrodes would be gradually eaten away as current passes or chemical substances may be liberated at the electrodes either as deposits or as gases. Thus chemical changes occur at the point where the current changes from a conduction current of electrons to a convection current involving ions. This process is called *electrolysis* and has considerable industrial importance. It is also important because it limits the amount of direct current that can be applied to the tissues.

The process of electrolysis was originally discovered by an anatomist, Sir Anthony Carlisle, who, with William Nicholson, showed that water could be decomposed to hydrogen and oxygen by means of an electric current (see below). Subsequently Michael Faraday was involved with a scholastic friend in coining the words related to this process; thus electrolysis, electrolyte, electrode, anode and cathode are still in use. These two latter words are from Greek, *anodos* meaning 'a way up' and *kathodos* 'a way down', referring to the current direction.

The electrolysis of water illustrates what happens. The presence of ions other than those of water increases the effect by providing many more ions. The electrolysis of water with a little sulphuric acid is usually described, although only the water is decomposed. The electrolyte will contain hydrogen ions (H^+) and hydroxyl ions (OH^-) produced by the dissociation of water, and $2H^+$ with each SO_4^{2-} from the sulphuric acid. When the electrodes are charged, H^+ migrate to the negative electrode (cathode) where they receive electrons from the metal and so become neutral hydrogen atoms. Pairs of hydrogen atoms combine to form molecules and can be given off as gas. In summary:

$$H^+ + \text{an electron} \rightarrow H$$
$$H + H = H_2$$

The OH^- and SO_4^{2-} similarly move to the positive electrode (anode) but only the OH^- give up their electrons; the SO_4^{2-} remain in solution, a characteristic called *preferential discharge of ions*. As the OH^- lose electrons and become neutral hydroxyl atoms they combine in pairs to form water molecules and oxygen atoms. These oxygen atoms also combine in pairs to form molecules of oxygen gas. In summary:

$$OH^- - \text{an electron} \rightarrow OH$$

$$OH + OH \rightarrow H_2O + O$$

$$O + O \rightarrow O_2$$

Thus electrons have been added from the metal electrode to the solution at the cathode and removed from the solution on to the metal plate at the anode, effecting the continuous passage of electrons from negatively charged cathode to anode. From the outside it appears that there is a steady flow of electrons through the electrolyte from negative to positive but in reality the process is a two-way movement of charges, ions, in the solution.

A similar process occurs in salt solutions. Sodium chloride dissociates to Na^+ and Cl^- which act in the same way as the sulphuric acid, supplying plenty of ions but being little involved due to the preferential discharge of ions. A similar result occurs with all acids, bases and salts in electrolytic solutions, including the complex mixtures of many different ions in solution which form the body tissue fluid.

A consequence of what has been described is the accumulation of H^+ in the region of the anode due to the constant removal of OH^-, hence the region becomes acidic. At the cathode the removal of H^+ leads to the opposite effect. This leads ultimately to damage if it is allowed to occur in living tissue. It may also have therapeutically beneficial effects if the effects are limited to mild irritation (see *Electrotherapy Explained*, page 17).

This, then, is the way in which currents pass in the body tissues and the changes that will occur at the junction of the conducting electrode and the electrolyte. It should be noted that when current is passed through the tissues via an electrode and saline-soaked pad or gel, the changes occur at the junction of electrode and pad. It may take some time before the solution becomes sufficiently alkaline to cause tissue damage. Since there tends to be rather more free alkali formed than acid, damage is more likely to occur at the cathode.

It should also be understood that effects at the tissue–electrode junction are dependent on the current density, i.e. current per unit area, and on the time for which the current passes. If very short pulses of current are applied there may be sufficient time between pulses for dissipation of all the chemicals. Naturally if the current direction is regularly reversed, as occurs in many therapeutic current applications, the effects will be entirely neutralized.

CURRENTS IN SEMICONDUCTORS

Consideration has so far only been given to the conduction properties of solids in which currents are movements of electrons in conducting materials, mainly metals, and non-conductors, insulators, in which easy electron movement from atom to atom is not possible in normal circumstances. The reason for these different characteristics is that, although all electrons in any given material are involved in the cohesive binding forces which hold the atoms, molecules and crystals together, some have only a negligible part in the binding processes so that they can move freely in the material without altering its physical properties. This occurs in metal in which some outer orbit electrons are free to move from atom to atom and hence provide a movement of charges – an electric current. Naturally, to effect any electron movement energy must be added to the material. An electric field will cause electrons to drift in the appropriate direction and heating the material will increase the random electron motion (interfering with any steady drift and hence causing greater resistance to currents in metal conductors; see page 41). In metals at room temperature there is sufficient energy available for much random electron movement, thus a piece of metal can be pictured as a set of vibrating but none the less fixed atoms in a sea of randomly moving electrons. In insulators the atoms are vibrating but it takes a great deal of energy to cause electrons to move from one atom to another and, if they do, it alters the structure or nature of the material. The distinction between the two is therefore that some electrons in con- ductors are able to accept the small amounts of energy available from an electric field to alter their energy level and move out of position whereas electrons in insulators need vastly more energy to reach the necessary energy level for movement within the material. However there are some materials which exhibit the characteristics of an insulator but have some electrons which can behave as electrons in a conductor. Not unreas- onably such materials are called *semiconductors* but it should be realized that they are not simply high-resistance conductors or low-resistance insulators but materials whose conducting ability alters under different circumstances.

Intrinsic semiconductors have a small number of free electrons at room temperature, i.e. at this temperature some electrons are able to move freely through the material, allowing it to behave as a conductor. Clearly these conducting properties are temperature-dependent. Ger- manium is an example of an intrinsic semiconductor.

Impurity semiconductors achieve the same characteristics by the addition of very small quantities of some other element. Thus adding a few parts per million of phosphorus to pure silicon converts it into an impurity semiconductor – a process called *doping*. In the same way pure germanium can be doped, increasing its usefulness as a semiconductor. Both germanium and silicon, in common with carbon, have four electrons in the outer orbital; they are tetravalent. This dictates the way in which they form crystalline structures. Each of the four electrons takes a part in binding the atom into a crystalline lattice with its fellows. If a small quantity of impurity is added the atoms of the impurity are scattered amongst the majority atoms fitting into the crystal lattice arrangement. Now, if the impurity has five outer orbit electrons, like phosphorus or antimony, four of them are involved in the crystal bonding arrangement but the fifth is free. If there were no thermal energy, i.e. at 0 K, it would remain weakly attached to its parent atom but under normal circumstances it wanders freely through the crystal- line structure. All electrons are identical so that the freely wandering electron can swop places with an electron fixed in the crystalline lattice, leaving that electron free.

If an atom with three outer electrons, boron or indium for example, is introduced as an impurity into the lattice structure these three electrons are locked into the lattice but there is a space or hole left in which an electron can readily fit. Thermal energy enables a nearby electron to fill the space so that in effect the hole moves through the material. The moving electron is a negative charge but the moving hole is, in effect, a positive charge. Impurity semiconductors are thus of two kinds: those in which electrons are the moving charges, called *n-type* because negative charges are moving, and those in which the positively charged holes move, hence they are called *p-type* (Fig. 4.40). The n-type is said to have a donor impurity, because it gives electrons; the p-type has an acceptor impurity because it accepts electrons. The n-type semiconductor is like a typical metal conductor in which electrons wander randomly through the material, whereas in the p-type it is the absence of an electron which 'moves' through the material, produced by short-range electron move- ments. If an electric field is applied to either type a current will pass. In the case of an n-type, a battery adds electrons from its negative terminal, which displace freely wandering electrons towards the positive terminal which accepts them. In the case of the p-type, electrons are withdrawn at the positive terminal, forming holes which are able to move towards the negative terminal and accept electrons as they arrive there. It must be understood that the n- and p-type semiconductors have no charge – they are electrically neutral as bulk material – until the battery or other potential is applied.

The p–n junction

If a piece of n-type semiconductor is in contact with a piece of p-type, electrons diffuse randomly across the junction from n- to p-type and holes diffuse in the opposite direction. Close to the junction the holes on the p-type side become occupied by electrons from the n-type and vice versa. This leads to a narrow *depletion layer* on each side of the junction in which there are very few free-charge carriers left. Thus the n-type side

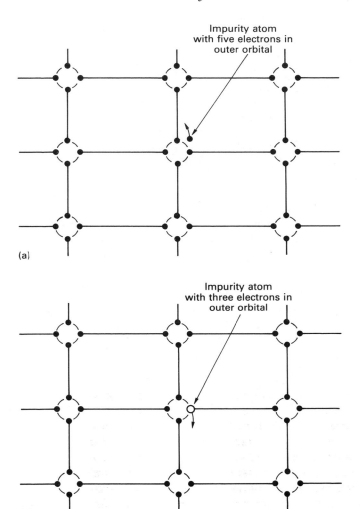

Fig. 4.40 Impurity semiconductors.

has lost electrons and so becomes positively charged and the p-type side has gained them and so becomes negatively charged. This forms a barrier or junction potential and restricts further movement of charges. This situation is illustrated in Figure 4.41.

Now, if a battery is connected across the p–n junction in such a way as to make the p-type side more negative still and the n-type side more positive, it will cause an increase in the barrier potential; that is, it will increase the resistance of the junction so that no current can flow, see Fig. 4.42a. (Actually a small current may flow due to some electrons being released by thermal energy but this will also depend on the potential difference applied.) If a battery is connected so that the barrier

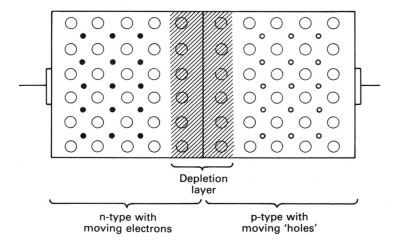

Fig. 4.41 A p–n junction.

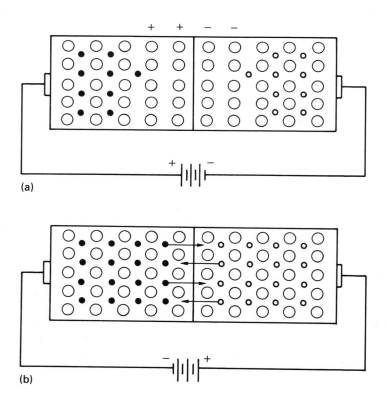

Fig. 4.42 A junction diode: (a) no current flowing; (b) current flowing. In this situation the barrier potential is much reduced, allowing electrons to move across the junction from left to right and holes to pass across in the opposite direction.

potential is reduced, that is, the n-type side is made negative and the p-type side positive, the effective resistance is much reduced so that a large current exists. Continually adding electrons at the n-type end and removing them at the p-type end creates a flow of charges just like that in a metal conductor, as shown in Figure 4.42b. This device, *a junction-diode* as it is called, will allow current to pass in one direction only and is the basis of all solid-state electronic devices.

At room temperatures the randomly moving electrons in the n-type material will have a range of energies, a few having high levels and a large number low levels, similarly for the holes in the p-type material. Consequently when a voltage is applied in the forward direction across a p–n junction some of the energy is used initially to free charge carriers so that the graph of current against voltage (Fig. 4.43) is exponential at first. Ohm's law is obeyed but the resistance alters with applied voltage. In a silicon diode virtually no current can pass in the reverse direction until the insulating properties of the depletion layer break down, which can be around 1000 V. (Germanium diodes tend to break down with much lower reverse voltages and are more susceptible to damage by temperature changes; hence silicon solid-state circuitry is more widely used now.)

Rectifier function

Because the junction diode allows current to flow in one direction only it can act as a *rectifier*, that is it converts a.c. to d.c. This can be done by placing a suitably sized junction in series with an a.c. circuit. This would lead to half-wave rectification, i.e. a series of pulses of current as the junction allowed the a.c. to pass in one direction but not the other. To achieve full-wave rectification a group of four junctions is used in one encapsulated unit. The arrangement is shown in Figure 4.44 which also shows the customary electrical symbol for a junction diode. It is very important to realize that the direction of the arrowhead in this symbol is the direction of conventional current, that is, opposite to the flow of electrons.

The bridge arrangement works as follows. During each half-cycle of a.c. one mains terminal is positive and the other negative. This reverses during the other half-cycle. Suppose the upper mains terminal is positive and the lower negative – conventional current can flow through B, around the working circuit and return through C to the negative lower mains terminal. The movement of electrons is actually in the opposite direction, through D, the working circuit and A. During the second half of the cycle when the upper mains terminal is negative and the lower positive, conventional current can pass through D, around the working circuit in the same direction as before and return through A. Hence current in the working circuit is unidirectional. It will, of course, be varying since the mains is a sinusoidal current. If it is necessary to

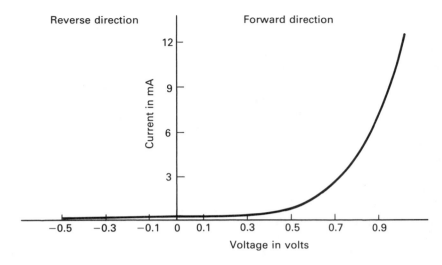

Fig. 4.43 A graph of current against voltage when a voltage is applied in a forward direction across a p–n junction.

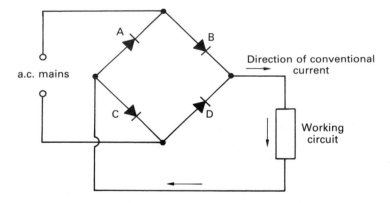

Fig. 4.44 A four-way rectifier circuit for full-wave rectification. The symbol —▶|— (which can be shown —▷|—) indicates a diode allowing current in one direction only. Note that the direction of the arrowhead shows direction of conventional current – electron flow would be in the opposite direction.

have a steady d.c., suitable capacitors and an inductance would be added to smooth the current.

Photodiodes

If a potential difference is applied to a junction diode in the reverse direction, electron–hole pairs are spontaneously formed in the depletion layer. This gives rise to a small reverse current. Radiations with sufficient photon energy can create extra electron–hole pairs and so increase the reverse current. This provides a means of measuring light and ultraviolet radiation.

Light-dependent resistors

These are also known as photoconductive cells and are specialized semiconductors whose resistance alters markedly when light is applied. In this way light can be used to switch other devices on or off.

Thermistors

Thermistors are semiconductor devices whose resistance is temperature-dependent. Resistance decreases exponentially with increasing temperature as electrons and holes are freed by heat. These devices are quite sensitive and accurate and can be made to be very small. They are often used in skin thermometers as a little bead placed on the skin surface as well as in small probes to be inserted into the tissues or body cavities (see Chapter 7).

Light-emitting diodes

A light-emitting diode (LED) is a special kind of p–n junction made from different materials which emits photons when electrons combine with holes in the junction layer. The result is the emission of a little light by the material which can be made into any shape. These are familiar as the figures and displays on clocks and calculators. An arrangement of seven bar-shaped LEDs is used to represent each digit by causing the appropriate bars to light up. In a similar way carefully constructed semiconductors can be made to emit visible or infrared radiations of a single wavelength, providing a laser source (see Chapter 8).

Transistors

These are semiconductors with three terminals. They are the modern equivalent of triode valves. They are of two kinds – junction transistors and field effect transistors.

The *junction transistor* consists of two p–n junctions set closely together so that the flow of current across one junction affects the current through the other. The two junctions include three separate pieces of doped semiconductor and can be either a p–n–p or an n–p–n transistor. The connections, which are called emitter, base and collector, are shown with their symbols in Figure 4.45. Note that the arrowhead on the emitter base shows the direction of conventional current; electron flow is in the opposite direction. Adhering to the convention allows the two types of junction transistor to be distinguished simply by the direction of the arrow, which is always placed on the emitter–base pathway. At its simplest level we can say that the base current directly affects the resistance between the collector and emitter. As it increases, resistance decreases, with the result that the collector current is proportional to the base current. When there is no base current no collector current flows. In general the collector current is very much greater than the base current. In this way the transistor can act as a current amplifier. Current gain/amplification factor = collector current divided by base current and can be 50–500. The fact that the resistance can vary provides the name transresistor or transistor.

A p–n–p transistor would work in the same way but holes rather than electrons form the main current carriers.

The *field effect transistor* consists of a narrow channel of one kind of doped semiconductor diffused on to a small piece of opposite type. These devices are often called *mosfets* – *m*etal *o*xide *s*emiconductor *f*ield *e*ffect *t*ransistors. In principle these devices work by regulating the charge carriers in the channel by an applied potential difference. They are also used in amplifying circuits.

Integrated circuits

The way in which semiconductor devices are manufactured allows many of them to be made at once on a single piece of silicon. Several thousand identical transistors can be made on a single, 5 cm wide sheet

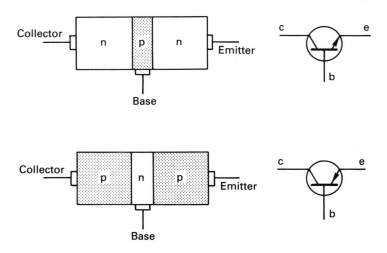

Fig. 4.45 Junction transistors and their symbols.

of silicon and then cut apart. This allows them to be made cheaply. A similar process allows many different components, transistors, diodes, capacitors, resistances etc. to be constructed on the same piece of silicon, called a *chip*. This enables complete circuits to be assembled on a single silicon chip, hence called integrated circuits. These can also be made very small indeed, some about 1 cm^2. They may take a long time to manufacture but as many identical circuits can be made simultaneously they are relatively cheap.

These integrated circuits are of many different kinds: complete amplifier circuits of various sorts can be bought – oscillators, timers and many others. Most modern electronic devices are built of a series of such integrated circuits connected to the other operating components. In fact, virtually complete circuits for pocket calculators or small radios can be made on a single chip. Most modern therapeutic electrical stimulators have oscillator and timing circuits based on a single chip.

THERMIONIC EMISSION AND SPACE CHARGE

The movement of electrons in conductors and semiconductors has been considered in the preceding pages but streams of electrons can cross empty space existing as a current in space. This leads to interesting effects and important applications.

When a metal is heated the random movement of its constituent electrons increases, as has been discussed already. Electrons near the surface of the metal, at the boundary with air, are also excited and some will be flung off the surface of the metal. As this leaves the metal positively charged, the electrons are attracted back. The extent to which they are thrown off is also constrained by the presence of air molecules. None the less at any point in time there will be a number of electrons free from the metal surface – an *electron cloud* – which will enlarge with greater heating. This is called *thermionic emission*. The situation is analogous to that of a juggler continually tossing up and catching balls. At any time a photograph would show some objects in the air but each is continually being thrown up and falling down; the more energetic the juggler, the more objects in the air at a time and the higher they are thrown.

Now, if the metal is given a negative charge and air molecules are removed, the electron cloud or *space charge* will be larger and if a positively charged piece of metal were brought close, some electrons would be attracted to it. Thus a flow of electrons can be made to occur in space.

Although this effect was first noticed by Edison (1847–1931), it was Fleming (1849–1945) who devised the first thermionic diode valve in 1904. Two electrodes are sealed into an evacuated glass bulb. One, heated by a separate circuit, is negatively charged and called the cathode, while the other is connected to the positive side of the circuit and called the anode. The space charge developed around the cathode allows electrons to flow to the anode so that the device conducts a current. If the heated electrode is made positive and the other electrode negative no current can pass because the space charge is made much smaller. The valve therefore allows a current to flow in one direction only.

By adding a third electrode between the cathode and anode the current through the valve can be controlled by quite small voltage changes in another circuit. The development of this triode valve was due to the American Lee de Forest in 1906. The third electrode is in the form of a mesh or grid of wire. The triode can act as an electronic switch, very rapidly turning current on and off, and as an amplifier, allowing a small voltage change to give a much larger but identical change. Figure 4.46 shows the symbols for the diode and triode valves with an illustrating circuit for the triode. It is customary to ignore the separate low-voltage circuit needed to heat the cathode in such diagrams. If there is no charge on the grid, or only a very small negative one, electrons can flow from cathode to anode through the holes in the grid. If the grid is made slightly more negative than the cathode it repels electrons, allowing fewer to pass through to the anode, so reducing the current across the valve. If the grid is made still more negative it will completely stop the flow of electrons, thus stopping the current. Because the grid is situated closer to the cathode than the anode it has a much greater effect, consequently small changes of grid voltage will be able to control very much larger voltages between anode and cathode. A change of, say, 2 V on the grid could cause a 100 V change across the resistance, called the anode load in Figure 4.46. In this example it is suggested that 2 V on the grid completely stops the anode current, as it is called, so that no matter how much more negative the grid is made, no further change occurs; 2 V negative is the cut-off point. However, changes in grid voltage between −2 V and zero (zero voltage with respect to the cathode) cause parallel but much greater changes in voltage across the anode load. This is the amplifying function of the triode valve. Small voltage signals applied to the grid, such as the spike potentials generated in contracting muscles, can be amplified to produce a large strong signal sufficient to cause a sound in a loudspeaker or a mark on a chart recorder, producing a typical electromyographic pattern.

The cathodes of diode and triode valves are coated with special materials which give good thermionic emission at relatively low temperatures. Modifications of the triode valve, in particular to enable it to amplify high-frequency signals, involve an additional electrode, the screen grid, forming a tetrode and sometimes a fifth electrode, the suppressor grid forming a pentode valve. The electrodes are arranged inside an evacuated glass bulb so that the heated cathode is surrounded by the grid which is itself surrounded by the anode which is a metal cylinder.

While the invention of the diode and triode valves can be said to have initiated the electronic era, they have been replaced by semiconductor

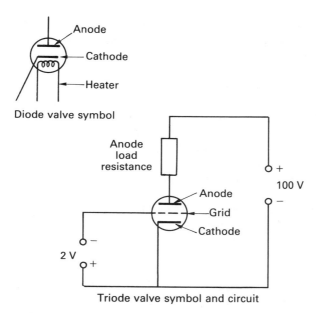

Diode valve symbol

Triode valve symbol and circuit

Fig. 4.46 Valve symbols and circuits.

p–n junctions and transistors. The rectifying action of p–n junctions was originally the province of diode valves and in some special circumstances, involving large currents, these valves are still used. The amplifying and switching functions of the triode are now done by transistors. The advantages of semiconductor devices over valves are evident in that the former are smaller, cheaper to manufacture, more robust and use less power since they do not have heater circuits.

Electron beams

By causing electrons to accelerate in space in an evacuated bulb it is possible to cause a stream of them to miss the anode and strike the glass envelope. On doing so they cause fluorescence and because they travel in straight lines, like radiation, away from the cathode they were originally called *cathode rays*, a name which has been retained. There are many electron beam or electron gun devices – the cathode ray oscilloscope, the electron microscope and, most familiar of all, the television tube.

A cathode ray oscilloscope consists of a heated cathode and an anode with a hole in the middle, all contained in an evacuated glass tube. Actually several specially-shaped anodes and other electrodes are used in practical cathode ray tubes in order to produce a narrow beam or stream of electrons. The principle is the same as a simple diode valve; the heated, negatively charged cathode gives off electrons by thermionic emission which are accelerated towards the anode which carries a strong positive charge. The acceleration causes some of the electrons to pass straight through the hole in the anode to strike the glass end of the tube. This surface is coated on the inside with a fluorescent material which emits light when struck by electrons. The electrons return via a conducting material such as graphite on the inside of the tube. Thus a spot of light is seen on the end of the tube (Fig. 4.47). The beam of electrons, moving negative charges, will be influenced by any close electric field. If the stream of electrons is made to pass between a pair of metal plates carrying opposite charges the electrons will be attracted towards the positive and repelled from the negatively charged plate. The beam of electrons is thus bent to a degree proportional to the charges on the plates. If the plates are sited above and below the beam it will be moved up and down. If an additional similar pair of plates is placed on each side the beam could be moved sideways (Fig. 4.47). With an arrangement like this, the electron beam, and hence the illuminated spot, can be moved about on the end of the tube to any desired position. With suitable circuitry the spot of light will trace a graph of voltage variations against time and thus show visually the shape of any electric pulses applied to the oscilloscope. Such oscilloscopes are common pieces of laboratory equipment and are sometimes incorporated into electrotherapeutic apparatus.

A television tube is an extension of the cathode ray oscilloscope in that the light spot is made to move from side to side and flick back rapidly to produce another line just below the first. This process continues to cover the whole screen with 625 lines in a very short time (about 40 ms) and

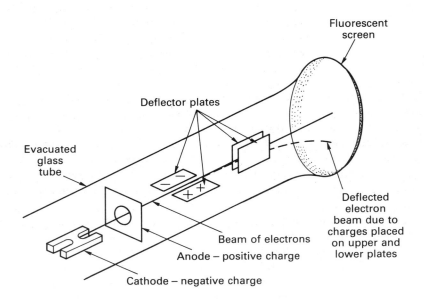

Fig. 4.47 A cathode ray tube.

the intensity of the electron beam is varied by the incoming signal to give a continually varying spot of light and hence build up a picture. The fluorescent screen and the phenomenon of visual persistence ensure that a continuous apparently steady picture is perceived. Colour television involves three separate electron beam emitters and a special fluorescent screen covered with little spots which emit appropriately red, green or blue light when struck by electrons.

The control of electron beam movement has been described as being due to electric charges on the deflector plates but the same effect can be achieved with magnetic fields which are used in television sets. It will be recalled (page 72) that magnetic force will provoke electron movement at right angles to the direction of the magnetic field. Thus the electron beams can be controlled by strong electromagnets placed outside the tube – variations of magnetic field acting in a similar way to the charges on the deflector plates. The ability to deflect or bend beams of electrons from an electron gun device with electric or magnetic fields allows the focusing of such a beam in the same way as beams of light can be focused. Such electron lens systems can be made to form an electron microscope on the same principles as an ordinary light microscope. In a transmission electron microscope the electron beam is passed through a very thin slice of the object – human tissue, for example – and some electron scattering occurs when denser parts of, say, cells are met. The resulting electron beam then passes through a magnetic or electric focusing system to produce a very much enlarged image on a fluorescent screen which is usually photographed and hence called an electron micrograph. In a scanning electron microscope the surface of the object is scanned by the electron beam and emitted electrons are focused on to a screen, giving a three-dimensional image. The great advantage of electron microscopes over ordinary light microscopes is that they have much greater resolving power because they are not limited by the wavelength of light (see Chapter 8). In fact electron microscopes can give magnifications of well over 100 000 times and achieve resolutions down to about 1 nm. They are, however, expensive and difficult to operate because the electron beams and the specimens must be kept in a near vacuum.

When very high potential differences are applied across the electrodes of a cathode ray tube, X-rays can be produced (see Chapter 8). This was discovered by Wilhelm Röntgen in 1895. They are produced when a stream of electrons strike matter and are therefore emitted at the anode of the cathode ray tube, provided a high enough voltage (100 kV or more) is employed. The properties of X-rays are considered on page 211.

5. *Production of currents for electrotherapy*

Functional components
  Power source block
  Oscillator circuit
  Production of shaped pulses
  Modification of electric pulses
  Amplification and depolarization of electrical pulses
  Control
  Measurement of output
  Output to the patient
  High-frequency currents
  Pulsed shortwave sources

From what has been described and explained so far, it is possible to understand how electrical circuits can be devised to deliver voltages and drive currents of any kind. In spite of the complexities, already and about to be considered, it must be realized that electrical currents of all kinds only differ from one another in three ways:

1. Their magnitude (intensity) in amps or milliamps.
2. The time for which they pass.
3. Their direction.

The rate of current change is simply the magnitude changing with time. The description of therapeutic currents is hampered by the fact that many such currents have a historical basis and naming was often connected with the way they could be produced at the time. For example, *faradic current* was originally generated by an induction coil, as considered on page 86. Likewise, current pulses of certain long durations were known as interrupted direct current because that is the way in which they were produced.

The nature of therapeutic currents and their physiological and therapeutic effects are considered in *Electrotherapy Explained*, Chapters 2 and 3. It is intended to elucidate here the principles on which electrotherapeutic sources operate, as an understanding of what is being used is an essential part of proper professional and rational therapy. It is not intended to provide detailed descriptions and technical data of particular electrotherapeutic sources.

FUNCTIONAL COMPONENTS

All electrotherapeutic apparatus can be described in simple terms by a series of functional components. These can be shown in a simple block diagram (Fig. 5.1). These blocks illustrate a group of components which perform the particular function indicated. In many modern devices the blocks are often manufactured as an integrated circuit, described on page 108, and are no longer a collection of discrete components connected together. The block diagram of Figure 5.1 is intended to illustrate a typical therapeutic electrical pulse generator and is not a specific machine. Similar diagrams illustrate other devices such as that for a shortwave diathermy source (see Fig. 5.9).

Power source block

Power may be drawn from the mains or from a battery. For small portable devices, such as transcutaneous electrical nerve stimulator (TENS) machines, batteries are appropriate. As described on page 94, zinc-carbon cells give 1.5 V each but by putting several cells together, forming a battery, larger voltages can be provided. Since many TENS devices will operate on quite small currents, the size of cell used to operate pen torches or calculators is sufficient to provide daily operation for some weeks. In some batteries the individual zinc-carbon cells are stacked together in the same casing. For example, the widely-used

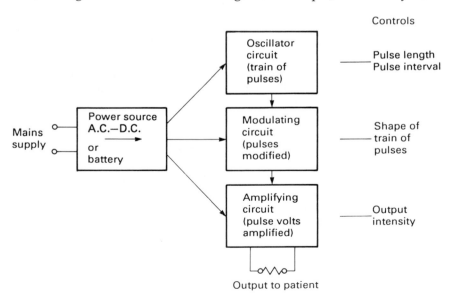

Fig. 5.1 Block diagram to illustrate electrical pulse generators.

thumb-size 9 V battery has six cells. Some use nickel-cadmium cells (see page 96). Batteries will provide direct current at a more or less constant voltage, although this may fall as the battery deteriorates. Some devices contain a circuit to give a warning light when the voltage starts to fall.

Power drawn from the mains will have to be modified. To reduce the voltage to a suitable level a transformer is used, as described on page 83. In the past sinusoidal currents were used for therapy; this was simply the mains current at reduced voltage. Similarly, diadynamic currents are mains-type current at reduced voltage with half-wave rectification (see page 105 and *Electrotherapy Explained*, Chapter 3). For most therapeutic stimulators constant direct current of a set voltage is derived from the mains and subsequently modified. Therefore the power source block or power supply unit consists of a transformer to reduce the voltage, a rectifier circuit to give full-wave rectification, such as the bridge circuit shown in Figure 4.44, and a smoothing circuit to remove the regular variation of the rectified alternating current. This latter often consists of a capacitor in parallel and a series inductance to smooth variations in voltage and flow respectively. It will be recalled that a capacitor will charge with an increase of potential difference and discharge with a fall of potential difference across it. An inductance resists both a rise and a fall of current by self-induction. Thus both lead to a smooth flow. There may be further electronic circuitry to ensure a very precise voltage and completely smooth current.

If direct current is required, for iontophoresis for example, then the addition of a control (potential divider) and a milliammeter to measure the current through the patient are all that is required (see Fig. 5.8). The only other components of this circuit are an on/off switch and a light to indicate that the power source circuit is operating. As already noted, such mains-modifying circuits are nowadays manufactured as a single integrated circuit which is supplied as a unit to change the mains to a certain precisely specified voltage – 240 V rms mains to 12 V d.c., for example.

Oscillator circuit

The hearts of many electrotherapeutic devices are circuits to produce oscillating or alternating currents of a particular frequency. There are several ways of producing electrical oscillations. A simple and widely-used circuit consists of a capacitor and inductance (Fig. 5.2) in what is called a parallel resonant circuit. On page 7 the motion of a simple pendulum was considered, producing a regular change from potential energy to kinetic energy. The natural frequency of swing is dependent on the length of the pendulum, i.e. the size of the system. If a pendulum is allowed to swing in air the amplitude – but not frequency – of the oscillation will gradually diminish as energy is lost until the pendulum comes to rest (Fig. 5.3a).

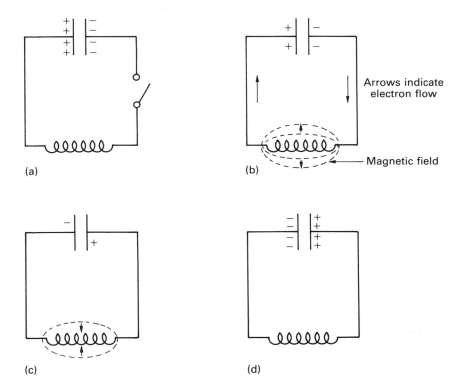

Fig. 5.2 Electrical oscillating circuit.

The electrical oscillating circuit of Figure 5.2 works in a similar way. There is a continuous to-and-fro movement of electrons around the circuit with a regular change from potential energy, when the capacitor is charged, to kinetic energy while current is passing in the inductance. Consider Figure 5.2. In (a) the capacitor is charged by some outside force so that the left-hand plate is positive and the right-hand plate negative. As the circuit is open no current can pass, so that the deficiency and excess of electrons on the left- and right-hand plates respectively remain as a store of energy – potential energy. In (b) the circuit is closed and the capacitor discharges around the circuit. Current in the inductance rises and with it the consequent magnetic field. This rising magnetic field cuts the turns of the coil, causing a back e.m.f. to act in the inductance thus diminishing the rate at which the current rises to maximum. (See page 75 for discussion on magnetic field and self-induction.) Energy is being stored in the magnetic field in the same way as the energy of motion (momentum) of the pendulum is a store of energy. As the current is rising the charge on the plates of the capacitor is falling as electrons leave the right-hand plate and are added to the left-hand plate. There comes a point at which the rising current is 'matched' by the falling voltage and the current then starts to fall, following the falling voltage.

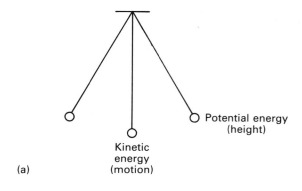

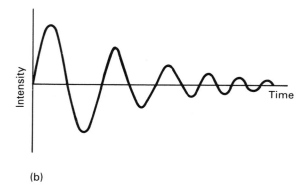

Fig. 5.3 The swinging pendulum and its exponential loss of amplitude with time.

As the current is now falling the magnetic field decreases, again cutting the turns of the inductance and causing a forward e.m.f., which prolongs the current flow, returning the energy that was temporarily stored in the magnetic field. Figure 5.2c illustrates the situation as the current is beginning to recharge the capacitor, forcing electrons on to the left-hand and off the right-hand plates. It is the equivalent of the swinging pendulum having passed its nadir, starting to climb up the opposite side by reason of its momentum. Ultimately the capacitor becomes fully recharged, but the opposite way, as illustrated in Figure 5.2d. Once again the charged capacitor has potential energy so that the whole process will be repeated, but in the opposite direction. Thus the capacitor discharges and charges repeatedly and a current rises and falls in the inductance with a consequent rising and falling magnetic field. The regular change from potential to kinetic energy occurs just as it happens with the swinging pendulum. For a further analogy of the system see Appendix C.

It is evident that the rate at which these events can occur will depend on the size of the capacitor and inductance; the smaller the capacitance and inductance of the circuit, the higher will be the natural frequency of oscillation, in the same way as a shorter pendulum will swing with a higher frequency. In fact the frequency (f) is inversely proportional to the square root of inductance (L) times capacitance (C) measured in henries and farads respectively:

$$f \propto \frac{1}{\sqrt{LC}}$$

To be more precise:

$$f = \frac{1}{2\pi\sqrt{LC}}$$

Thus, a circuit can be made that will generate oscillations at any required frequency by choosing an appropriately sized inductance and capacitance. The oscillations will be in the form of a sine wave, exactly like the trace of the swinging pendulum considered in Chapter 3 and shown in Figure 5.3b. In the same way, and for the same reasons, the amplitude will diminish in an exponential manner, as also illustrated in Figure 5.3b. At each oscillation some energy will be lost. In the case of a swinging pendulum this is principally due to friction with air molecules. In an electrical oscillating circuit the analogous loss would be heating of the inductor which must have some ohmic resistance in a real circuit so that some of the energy is converted into heat. In a real electrical circuit there will be other energy losses such as heating of the dielectric of the capacitor and losses from the varying magnetic field. Thus the electrical oscillations are damped. In order to maintain continuous oscillations the lost energy must be replaced. Further, when energy is deliberately removed from the oscillating circuit to provide the therapeutic current this too must be replaced.

This can be done in an electrical circuit by feeding bursts of energy in at exactly the right moment in the cycle to make good the losses. This is analogous to the widely recognized parental experience of pushing a young child on a swing. Pushes must be applied as the child is swinging away. If pushed at the wrong moment, while the swing is moving towards the pusher, the swing and child will come to rest. Thus the energetic parent is adding energy to the oscillating system of the swinging child which, to be effective, must be added at the right moment. The pushes would therefore be at a frequency which matches the frequency of the swinging child.

In the electrical circuit the push is given by a switch – either a transistor or a thermionic valve – which is coupled to the oscillator circuit so that current is added in time with the oscillations (Fig. 5.4). In the left-hand circuit, current through the valve is added to the oscillating circuit, adding electrons to the left-hand plate and pulling them off the

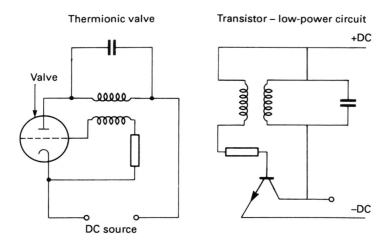

Fig. 5.4 Methods of boosting oscillations.

right-hand plate of the capacitor. As the capacitor discharges, driving electrons from left to right through the inductance, an opposite current is induced in the grid coil (by Lenz's law) to drive electrons on to the grid and switch the valve off. The two coils form a transformer. At the next oscillation, when current is passing in the opposite direction in the oscillator coil, it induces an opposite current in the grid coil to remove electrons from the gird, thus allowing a further burst of current. Thus a series of appropriate bursts of energy are added at the right moment in each cycle of oscillation to replace energy lost as heat or transmitted to another circuit. The maximum amount of charge added is, of course, limited by the size of the capacitor and voltage of the input source. In this way a continuous undamped, sinusoidal, oscillating current is maintained in spite of losses to other circuits.

The right-hand circuit using a transistor operates in a similar way. Current in the oscillator coil causes, by electromagnetic induction, a current in the opposite direction in the coil attached to the base of the transistor. It will be recalled from page 108 that altering the emitter–base current by quite small amounts can switch the collector current on and off. Thus the emitter–collector current through the transistor is switched on and off in time with the oscillations of the oscillator circuit, adding an appropriate burst of current each time it is switched on. There are numerous other ways in which the circuit may be made to give continuous oscillations using a transistor (or transistors) in other configurations.

Such circuits are used to produce high-frequency oscillations, for shortwave diathermy (SWD) for example (see later). The valve-operated system is found in older apparatus and is appropriate for high currents. The transistor-controlled version would produce low currents that would need amplification for therapeutic use.

Crystal oscillators are also much used, especially for circuits in which oscillations must maintain a very precise frequency. Capacitance-inductance circuits tend to drift very slowly, especially at higher frequencies. Quartz crystals have a piezoelectric property (see Chapter 6) which means that they alter their shape when a potential difference is applied across two opposite surfaces. This reverses when the electrical polarity is reversed. The crystal will exhibit mechanical oscillation at a frequency which is determined by its shape and size. When included in the circuit between a pair of electrodes it behaves like a tiny oscillating circuit, permitting only the frequency of electrical oscillation that matches its own natural resonant frequency. Such circuits are found in quartz clocks, watches and in many situations where precise timing is needed. These quartz oscillators will also, like the capacitance-inductance circuits, produce sinusoidal oscillations but there are many situations in which a train of pulses of regular form are needed which are of some other shape.

Production of shaped pulses

The production of a regular series of electrical pulses can be based on capacitance-resistance circuits, which were considered on page 55. It will be recalled that the capacitor in such a circuit can be made to charge or discharge over any chosen period of time, depending on the size of the capacitor and resistance, and that the discharge will be exponential. The analogy with a sand-glass egg timer was drawn. With such a circuit it is possible to set the precise length of time for an electric pulse and, with a second circuit, to set the length of interval between the pulses. What is needed is a means of switching the circuits on and off and this can be done by a transistor.

The way in which the transistor acts as an electronic switch is illustrated in Figure 5.5. When no base current is flowing the transistor is off in the sense that there is a high resistance between collector and emitter. If a small current is applied through the base it will decrease the resistance of the path between collector and emitter to allow a flow of current, turning the switch on. If the base current is made large enough it will reduce the resistance in the collector–emitter pathway to zero. Quite a small base current will have a large effect on the collector–emitter current so that small changes in base current can be made to turn the transistor completely off – acting as a switch.

By applying the output from one timing circuit with a transistor to another similar circuit and feeding back the output of that to control the first circuit, it is possible to produce a continuously oscillating current. Such a circuit is called an astable multivibrator or, more expressively, a flip-flop circuit (it flips from one state – a pulse of current – to the other state – no current). Figure 5.6 shows such a circuit.

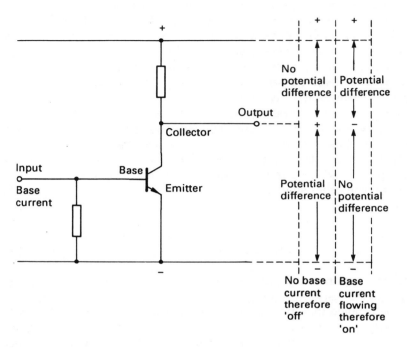

Fig. 5.5 The transistor as a switch.

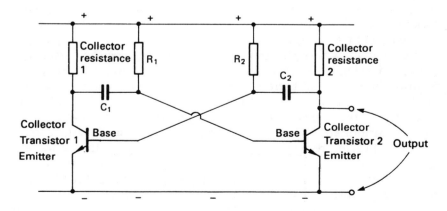

Fig. 5.6 An astable multivibrator circuit.

To see how this circuit works, assume that transistor 1 is not conducting because no base–emitter current is flowing, the base being negative. There is therefore no voltage drop across the collector resistance 1 so that the collector is at the voltage of the positive line. At the same time transistor 2 is on because the base is slightly positive, allowing a base–emitter current to flow. Thus current flows through the collector–emitter path of transistor 2, resulting in a potential difference across collector resistance 2 and hence the collector voltage will be negative. These assumptions are made to give a starting point in the cycle.

The capacitor C_2 now charges through the resistance R_2 at a rate proportional to their sizes. As the left-hand plate of the capacitor becomes positive, emitter–base current starts to flow in transistor 1 to turn it on. This allows collector current to flow so that the collector voltage becomes negative.

This negative voltage is transferred by capacitor C_1 to transistor 2 to turn it off. Capacitor C_1 will now charge through R_1 until transistor 2 is conducting again. Once this happens the process repeats, with transistor 2 causing a negative potential to be applied to the base of transistor 1 to turn it off again. Thus one transistor is fully conducting while the other is fully off and vice versa.

It can be seen that altering the resistance of R_1 and R_2 will alter the pulse intervals and pulse durations respectively. This can easily be done by means of a switch on the pulse generator which can be altered to select different predetermined resistances to give different predetermined times. Such a circuit would start to oscillate automatically once voltage is applied because the state of conduction of one or other transistor is bound to be slightly greater, thus turning the other off and starting the on–off cycle. The output across transistor 2 could have the form shown in Figure 5.7.

The flip-over from conducting to non-conducting and vice versa in these transistors can be very rapid, almost instantaneous, as previously explained. It is therefore justifiable to show pulses as square waves, as in Figure 5.7.

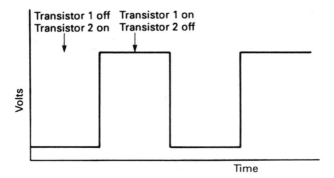

Fig. 5.7 Output from transistors shown in Figure 5.6.

Thus an astable multivibrator circuit can be made to provide a series of pulses of any desired pulse length and frequency. As noted earlier, modern apparatus uses integrated circuits and many different such circuits are produced, including versions of the circuit just described. There are many timer circuits available as integrated circuits to which appropriate capacitors and resistance can be added to give the required type and frequency of pulses.

Modification of electric pulses

Therapeutic currents are usually applied as a series of bursts of current or as a regularly rising and falling surged or ramped current, e.g. TENS or faradic-type pulses, as described in Chapter 3 of *Electrotherapy Explained*. The regular pulses produced by a multivibrator circuit just described can easily be modified to occur in a series of separate pulse trains or bursts by controlling it with the output of another separate multivibrator. To see how this might work, suppose that a multivibrator circuit set to deliver a series of 0.2 ms (200 μs) pulses at a frequency of 100 Hz – i.e. with pulse intervals of 9.8 ms – is applied across the transistor of Figure 5.5 in place of the supply. The second multivibrator is set to deliver a series of 2 second pulses separated by 3 second pulse intervals – i.e. a frequency of 0.2 Hz (or 12 per minute) – and this is applied to the base of the transistor of Figure 5.5. Thus during the 2 second pulse the transistor is 'on', so that a 2 second pulse train of 0.2 ms pulses at 100 Hz appears between the output and the positive line. For the following 3 seconds the transistor is 'off' and hence no output occurs. It can be seen that varying the resistances of the second multivibrator circuit can lead to many different lengths and frequencies of pulse trains. In circuits that use thermionic valves there is a similar arrangement in which triode valves perform the switching function of transistors for the multivibrator circuits and the voltage output of one can control the other by being applied to the grid of a triode valve.

There are various other ways in which trains of pulses can be generated. In many modern sources in which integrated circuits are utilized the multivibrators can be applied to a voltage-controlled amplifier, a further integrated circuit.

The train of pulses may need to be surged or ramped, that is to rise and fall gradually rather than instantaneously. This may be desirable to provide slowly rising and falling sensory stimulation in TENS or interferential therapy and a more physiological-like muscle contraction in response to muscle-stimulating currents. To achieve this the second multivibrator circuit is made to apply pulses that rise and fall at the appropriate slow rate. It can be seen how easily this can be achieved when it is recalled that the charging and discharging of a capacitor is exponential.

Amplification and depolarization of electrical pulses

It has already been noted that inexpensive, ready-made integrated circuits are built with transistors that operate with small currents at low voltages. For therapeutic currents rather higher voltages are needed – often tens of volts. A large range of possible applied voltages is needed because what matters therapeutically is the current density, i.e. the current per unit area, which depends on the size of the electrodes used for application (see *Electrotherapy Explained*, Chapter 3). The method of regulation is discussed below.

A circuit to increase the voltage and hence the current is called an amplifying circuit or amplifier. The transistor can act as an amplifier. In Figure 5.5 the circuit shows how a small varying current applied to the base of the transistor can lead to identical but much larger changes of collector current. By connecting the varying input via a capacitor, to insulate the transistor, and adding suitable resistances it is possible to produce a much larger output voltage. In modern circuitry cheap integrated circuits are available, which comprise several transistors, capacitors and resistors all in one tiny component, called an operational amplifier. In principle operational amplifiers are arranged to give very high voltage amplification of the order of a million times. This is called the gain of the amplifier. Since the actual amplification needed may be only about, perhaps, 10 or 20 times the original voltage for electrotherapy, the operational amplifier is fitted into a circuit with appropriate resistances that feed back some of the output signal to diminish the input partly and thus achieve the required level of amplification.

Of course, the additional energy needed is derived from the power source. The same end can be achieved with the use of a triode valve, the small changes in grid voltage causing large changes of voltage across the anode load.

Many electrotherapeutic sources are depolarized currents, that is, there are equal amounts of charge in each direction. This is because any d.c. component will cause chemical changes at the tissue–electrode junction (see page 100) which is not wanted with nerve-stimulating currents, although it may be appropriate for some other therapeutic currents (see *Electrotherapy Explained*, Chapters 2 and 3). Varying d.c. from the multivibrator circuits just described can easily be converted to a.c. by applying it to a capacitor. The capacitor will not pass a direct current – it will not allow the passage of electrons through the dielectric – but as the d.c. charge on one plate changes, say becomes more positive, it draws electrons on to the opposite plate, thus causing a flow of electrons in the direction of the capacitor. As the d.c. changes to become less positive, electrons will be pushed off the opposite plate, causing a flow away from the capacitor. Thus if the d.c. circuit varies regularly the output from the other plate of the capacitor will be evenly alternating. If the d.c. signal is irregular the alternating output will follow it and also be irregular.

Control

Thus currents have been generated, modified, amplified and depolarized to produce an appropriate electrotherapeutic current modality. Regulation and control at the various stages are also needed. It has already been explained how varying the resistance can alter pulse length and interval. Such variations can be set on the machine by altering a control knob or button which alters the particular resistance. The frequency, rate and length of surge, number of surges or bursts per minute can be controlled in a similar way. Timing of these and treatment length is achieved by a suitable timing circuit. Some machines provide automatic variations of current which can be preset or programmed in. Many interferential sources have arrangements which allow a regular change of beat frequencies to occur over a fixed time period (see *Electrotherapy Explained*, Chapter 3). The development of cheaper integrated circuits has allowed greater use of preset treatment arrangements (or treatment schedules) so that the source may be described as programmable or computer-controlled.

In most sources, indications of the type, frequency and intensity of applied current is usually provided by light-emitting diodes (LEDs; see page 107).

The most important regulation for therapy is that of the current to the patient. In many sources this is controlled by a variable resistance in parallel with the patient, as illustrated in Figure 5.8. This potential divider or parallel rheostat allows the potential difference applied to the patient to be zero and to be increased gradually thus ensuring that at 0 on the intensity control no current will pass through the patient but on turning the control a slowly increasing current can pass. Some electrotherapeutic sources have an electronic lock so that the output to the patient can only be switched on if the intensity control is at zero. This prevents a high output being suddenly applied to the patient if the machine is switched on with the intensity control left at high. Although some machines have methods of measuring the output intensity (see below), others simply rely on the position of the intensity control knob. This may have a numerical scale beside it but these are usually relative

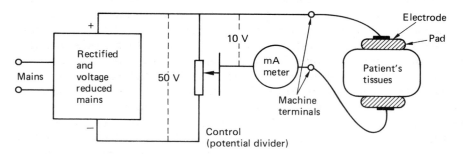

Fig. 5.8 Production of d.c. for patients.

indicators and represent no actual units of current or voltage. Further, some sources, especially small simple TENS units, have outputs which are not linear and do not follow the linear scale shown on the control knob.

Various methods and styles of control may be utilized, making the operation of machines superficially different. The output can be linked to the treatment timer circuit so that the length of treatment time must be specified before it can be applied.

Measurement of output

For adequate control some knowledge of the output to the patient is needed by the therapist. This can sometimes be provided by an analog meter (see page 69), showing the current or voltage being applied by a needle moving over a scale. Many modern sources display the same information in digital form. A set of LEDs and a suitable integrated circuit provide a numerical display. These are much easier to read accurately than an analog meter. It must be recognized that both kinds of meter usually measure the average or rms current of a rapidly varying therapeutic output. Other parameters are also often displayed digitally, such as the length of treatment time, surge time or pulse frequency.

Output to the patient

From what has been noted already, it must be evident that the current through the tissues from a therapeutic stimulator will vary with the impedance of the circuit through the patient. This will vary with the resistance of the leads and electrodes, the capacitive impedance of the electrodes and tissues (inductive impedance is usually negligible), the resistance of the tissues and the resistance of the electrode–tissue interfaces. This latter is often the most significant factor. These variations of total impedance will cause the current to vary. Stimulators are usually arranged as either *constant current* or *constant voltage* sources. This can be achieved by making the resistance of the circuit in the machine, in series with the patient circuit, so large that variations of the patient's circuit are negligible compared to the total impedance or by arranging a low internal parallel resistance. It can also be achieved by a special arrangement to monitor continually the impedance of the patient's circuit and feedback corrections to the output amplifier to ensure a constant output. Thus if the electrode–skin resistance rises a little due to drying out of the sponge or gel, the current is automatically increased to compensate. Such devices have current limiters fitted for obvious safety reasons. Much discussion has taken place over the therapeutic advantages and safety of constant current as compared to constant voltage pulses (Patterson, 1983). Constant current machines are considered preferable by some people if the electrodes will, or are likely to, be moved during treatment.

Discussion of the sources of electric currents for therapeutic purposes has so far assumed a single circuit, the principle illustrated in Figure 5.8, in which any effects on the tissues must occur in the pathway of current between the two electrodes and usually close to one of them. (See *Electrotherapy Explained*, Chapter 3, for a discussion of the physiological effects.) Interferential currents can be applied in this way but it is more usual to use two separate circuits for their application. These currents are of medium frequency, 4000 or 5000 Hz, generated in the way described above and of sine wave form. By applying two such currents in two separate circuits through the tissues and arranging a small frequency difference between them it is possible to achieve a combined low-frequency current as a consequence of superposition of the two wave forms (see Chapter 3). It is the equivalent of the beat frequency considered with reference to sound in Chapter 6. The resulting low frequency is the difference between the two medium frequencies; thus, if one circuit carries a 4000 Hz signal and the other a 4100 Hz signal, the low-frequency beat will be of 100 Hz. It is this low-frequency current that will trigger nerve impulses and hence have some therapeutic effect (see *Electrotherapy Explained*, Chapter 3). The frequency variations of one circuit will lead to changes of beat frequency and the incorporation of suitable timing circuits can allow this to occur regularly and automatically.

High-frequency currents

The very short pulse, or phase, of electrical energy of a medium-frequency current would need a high current intensity to produce a nerve impulse. The medium-frequency currents of 4000 Hz, just discussed, would have a pulse or phase duration of 0.125 ms (because each cycle has a peak in both directions). Still higher frequencies, hence shorter pulse duration, would need still higher intensities to cause nerve impulses (see *Electrotherapy Explained*, Chapter 3, for discussion on strength–duration curves). Thus if frequency of electrical stimulation is made very high, say in the megahertz region, then quite large currents can be passed without causing nerve stimulation. (The actual currents that might be used would depend on the size of the area to which they were applied, i.e. the current density would be the determining factor.) This would enable currents of the order of amperes rather than milliamperes to be passed in the tissues, which leads to heating. [By Joule's law, $H \propto I^2 R t$ where I (current) is in amperes, R (resistance) in ohms, t (time) in seconds and H (heat energy) in joules.]

Heat produced in the tissues by passing such currents was used in the past for therapeutic heating and was known as longwave diathermy. It is still sometimes used for tissue destruction during surgery. Such currents must be passed through the skin via electrodes and sponges or pads soaked in fluid containing ions. However, at these megahertz frequencies most of the energy is transmitted capacitively (see Chapter 4). At higher frequencies still, say around 30 MHz, the energy can be transmitted almost entirely capacitively so that there is no need for any wet pads. Currents will be passed if the tissues are made part of the dielectric of a capacitor, formed by the electrodes. Thus high-frequency currents are generated in the tissues in spite of being separated from the electrodes by insulation and an insulating air gap. These frequencies are in the shortwave radio band and hence therapeutic heating with such currents is called shortwave diathermy.

The principles on which shortwave diathermy sources operate are illustrated in the partial block diagram of Figure 5.9. Continuous high-frequency oscillations are produced either from a valve-controlled circuit, operating at a fairly high voltage (around 1–2 kV), which is provided by a step-up transformer from the mains, or by a transistor-controlled circuit operating at a much lower voltage but amplified before being applied to the patient; see Figure 5.4 which illustrates the two methods of producing oscillations.

Since oscillation frequencies are close to those of radio and television broadcasts, it is necessary to restrict those used for medical or scientific purposes to particular frequencies, so that they do not cause interference. Three frequencies have been assigned, by international agreement, and are shown in Table 5.1. The 27.12 MHz frequency is by far the most widely employed for therapeutic shortwave because it has the widest frequency tolerance, i.e. how far it is allowed to drift off the exact

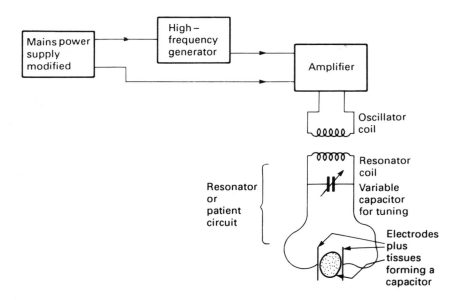

Fig. 5.9 Block diagram to show shortwave diathermy generation.

Table 5.1   Assigned frequencies and wavelengths

Frequency (MHz)	Wavelength (m)
13.56 ($\pm$ 6.25 kHz)	22.124
27.12 ($\pm$ 160 kHz)	11.062
40.68 ($\pm$ 20 kHz)	7.375

frequency. In order to maintain frequency stability and diminish radio interference, cavity (pot) resonators are used with the valve circuit. These are oscillating circuits of large capacitance, formed of a metal box of precise dimensions, enclosing the inductance. Frequency stability may be effected in the transistor controlled circuit by including a resonating crystal (see Chapter 6).

To transfer energy to the patient's tissues a separate circuit is used, shown as the resonator or patient circuit in Figure 5.9. The oscillator and resonator coils form a transformer so that electrical oscillations in the former will induce a similar oscillating current in the latter. If the two circuits have the same natural frequency, i.e. are in tune or in resonance, then there will be maximum transfer of energy. The oscillator frequency is fixed at 27.12 MHz but the natural frequency of the patient circuit will vary, depending on the nature and configuration of the tissues and electrodes. Since the oscillating frequency is proportional to

$$\frac{1}{2\pi\sqrt{LC}},$$

varying the inductance or capacitance will alter the frequency. In fact, a variable capacitor is used to tune the patient circuit so that it has the same natural frequency as the oscillator circuit under the particular load conditions. The tuning can be done manually, using the excursion of a meter or brightness of a light to indicate maximum resonance, or it is done automatically. In this case a motor driving the tuning capacitor is itself regulated by the output from the resonator circuit or the same effect can be achieved electronically. In this way it is able to maintain resonance even if the patient moves a little. The tissues may be coupled into the shortwave field, both as part of the dielectric of a capacitor, as shown in Figure 5.9, or as the load of an inductance, either in the form of a cable wound around or placed on the tissues or, more usually, fixed and mounted in a drum-type applicator. The tuning will work in the same way.

The output of the machine, the intensity, determines the heating in the tissues and may be regulated in several ways. The voltage applied to the oscillator coil can be varied in a series of steps by switching in different lengths of secondary winding of the mains transformer. On some machines, the coupling between the oscillator coil and the patient's circuit coil can be altered, either by physically moving one of the coils or by moving a screen between the coils. More modern machines regulate the output of the power amplifier by electronic means.

Pulsed shortwave sources

The shortwave sources already described can be modified to provide a pulsed output. The oscillator circuit is controlled by an astable multivibrator (see Chapter 4), which effectively switches it on and off to provide a series of short pulses of high-frequency (27.12 MHz) oscillations. The other parts of the circuit and the methods of coupling to the patient's tissues are the same as those already described. The pulse frequency and, on some machines, the pulse width can be varied. Pulse frequencies of between 15 and 800 Hz are available and pulse widths are 20 to 400 μs. The combination of shorter pulses of high-frequency energy at low pulse repetition frequencies produces no detectable tissue heating, hence should not be called diathermy. Longer pulses repeated more frequently can produce minimal, detectable heating. For a discussion on the nature and effects of pulsed shortwave, see *Electrotherapy Explained*, Chapter 10.

All shortwave sources have treatment timing circuits, sometimes connected to the on/off switches, to enable the precise treatment time to be preselected. Indicator lights for mains, treatment and sometimes tuning are also provided.

Ultrasound sources are described in Chapter 6: the oscillator circuits are similar to those already considered.

The foregoing was intended to describe the *principles* of machines used for electrotherapy. Consideration has been given to the sources that provide electrical currents of all kinds for therapy. Some machines provide one modality only but often different currents are grouped in various combinations in one piece of apparatus.

6.   *Sonic energy, sound and ultrasound*

Sonic energy is mechanical vibration causing the formation of longitudinal waves in matter. These waves and their propagation were discussed in Chapter 3 and reference should be made thereto, as necessary, throughout this section. This type of energy is most familiar as sound but the same waveforms occur at frequencies beyond the range of normal hearing and are therefore known as ultrasound or ultrasonic energy at the higher frequencies and infrasound at the lower. The word *sonic* arises from the Latin *sonus*, sound, and is used adjectivally to mean pertaining to sound waves – hence the use of terms like *insonation* for ultrasound therapy and *supersonic*. This latter can sometimes be synonymous with ultrasound as well as meaning faster than the speed of sound. The word sonar is an acronym of *sound na*vigation and *r*anging, referring to echo location.

Sound consists of those frequencies between about 20 and 20 000 Hz which can be heard by the human ear. This is an approximation since the limits of normal hearing cannot be exactly defined. Children may be able to hear frequencies somewhat greater than 20 000 Hz but this ability declines with age. Further, the frequencies that can be heard are strongly dependent on their amplitude (see later).

THE NATURE OF SONIC WAVES

As already noted, these are longitudinal waves; that is, there are variations of pressure that occur in the direction of travel of the wave. It will be recalled that all matter is made of discrete, randomly moving particles – atoms and molecules. In gases the movement is considerable but in liquids and solids it is more like a vibration (see Chapter 2). In all cases many millions of collisions occur between molecules every second. Consider the passage of a sound wave in air. Some object, such as a plucked guitar string, or the vibrating ruler described in Chapter 3, pushes the air molecules together at some point in front of it, causing a small region of compression – denser air. This compression affects adjacent air molecules ahead of it, causing them to be compressed. At the same time the vibrating force – guitar string or whatever – has started moving in the opposite direction, pulling apart the air molecules in front of it, thus causing a region of rarefaction – less dense air. Thus a region of higher pressure followed by one of lower pressure passes through the air to be followed by other identical waves as the vibration continues. What progresses through matter is the wave of pressure changes, not the individual molecules which remain, on average, in the same place. They are, of course, caused to move to and fro, in a regular manner, in the same direction as the wave. This is, of course, superimposed on their existing random motion. Thus the sound wave is a to-and-fro molecular displacement causing a pressure variation to occur in both time and space. Figure 6.1 is an attempt to illustrate this idea. The left-hand line of rectangles is intended to represent groups of molecules in one plane oscillating to and fro into the page and leading to a graph of compression/rarefaction. The same wave is shown at the two later times on the horizontal axis, illustrating that the pressure varies sinusoidally both with distance and with time.

For ordinary, everyday sounds the size of these pressure changes is quite astonishingly small, illustrating the efficiency of the ear–brain mechanism at detecting and recognizing these tiny variations. For normal conversation, the pressures above and below atmospheric pressure at the ear are about 0.01–0.03 Pa (pascal; $1\,\text{Pa} = 1\,\text{Nm}^{-2}$), which is only a few millionths of the atmospheric pressure. The actual molecular displacements in a sound wave of this size are correspondingly tiny. Detectable sounds can involve movements smaller than the diameter of an atom.

What has been said in describing a sound wave in air applies, in principle, to all sonic waves of whatever frequency or amplitude travelling in any medium – gases, liquids or solids. The characteristics and ways of defining waves are described in Chapter 3 (see also Fig. 6.1). The relationship, $v = f\lambda$, where v is velocity, f is frequency and λ is wavelength, is also discussed.

The size or amplitude of the sonic wave is the difference between the extreme position and the mean – thus, the greater the pressure above and below the mean, the greater the amplitude. The total energy carried by a wave of any given frequency will be directly related to its amplitude and frequency.

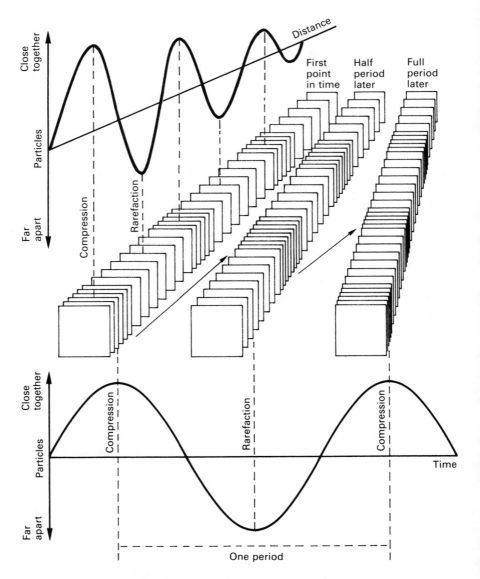

Fig. 6.1 Representation of a longitudinal wave.

VELOCITY OF SONIC WAVES

Longitudinal waves will pass through different media at velocities that are characteristic of the particular medium. This is because the velocity will depend on the distance between molecules and how easily they can move. Thus the velocity in solids and liquids is much greater than that in gases whose molecules must travel relatively greater distances before influencing other molecules. The speed of a mechanical wave in any material will depend on the elasticity and inertia of the molecules of the material. To be precise, the velocity (v) is equal to the square root of elasticity over density. For a solid:

$$v = \sqrt{\frac{\text{Young's modulus of elasticity}}{\text{density}}}$$

In liquids:

$$v = \sqrt{\frac{\text{Bulk modulus}}{\text{density}}}$$

For gases, the elasticity or springiness depends both on the pressure and on a number that varies from one gas to another. This number, which is actually the ratio of the specific heat of the gas at constant pressure to its specific heat at constant volume, is about 1.4 for air. Thus for air:

$$v = \sqrt{\frac{1.4 \times \text{pressure}}{\text{density}}}$$

Some examples of sonic wave velocities in different materials are shown in Table 6.1. Since, in a gas, any increase in pressure is an increase in mass per unit volume, i.e. density, changes in air pressure do not affect the speed of sound. However, if the density of air is decreased by a rise in temperature while the pressure remains constant, the ratio of pressure to density will increase and hence so will the velocity of sound. Thus, in summary, the velocity of sonic waves depends on the nature of the medium through which they are passing and is independent of frequency. The speed in air is independent of air pressure but proportional to the square root of temperature. The velocity of sound in air at 20°C is approximately 343 m s^{-1}.

INTENSITY OR MAGNITUDE OF SONIC WAVES

The amount of energy carried by a sonic wave of any given frequency will depend on the amplitude, as noted above. The intensity of the wave is the amount of energy that crosses unit area in unit time. This is measured in joules per square metre per second (J m^{-2} s^{-1}) or, since 1 W = 1 J s^{-1}, in watts per square metre (W m^{-2}). In many circumstances it is more convenient and usual to use watts per square cm.

Thus, sonic waves in any given medium can be fully described by frequency (hence wavelength; see Table 6.2) and intensity in watts per square cm. These are the dosage measurements customarily used in

Physical Principles Explained

Table 6.1 Some examples of approximate velocities of sonic waves

Material	Velocity (ms^{-1})
Solids	
Granite	6000
Glass	5000–6000
Iron	5000–6000
Bone	3445
Lead	2100
Rubber	1800
Tendon	1750
Cartilage	1665
Muscle	1552
Liquids	
Blood	1566
Seawater (at 25°C)	1531
Freshwater (at 25°C)	1498
Fat	1478
Gases	
Air (at 20°C)	343
Air (at 0°C)	331
Oxygen (at 0°C)	316
Carbon dioxide (at 0°C)	259

Figures from Cromer (1981) and Frizzell and Dunn (1982).

ultrasound therapy but it must be recognized that these describe the ultrasonic energy *applied* to the tissues. The energy *absorbed* in the tissues – the real dosage – is largely unknown. In fact, quantifying the dose is extremely difficult and in this respect the application of therapeutic ultrasound becomes something of an art – a matter of judgement based on a knowledge of the way ultrasonic energy is absorbed in the various tissues in the treatment field.

SOUND

As already noted, frequencies between 20 and 20 000 Hz are those that can be heard by the human ear and recognized as sound. Infrasound below 20 Hz is detected as vibration. Most normal speech and music occurs at frequencies of a few hundred to a few thousand hertz (Table

Table 6.2 Frequencies and wavelength of sound and ultrasound

Frequency (f) in hertz (Hz)	Wavelength (λ) in metres (m)	
Sound in air at velocity of 343 ms⁻¹		
20	17.15	Lowest frequency sound
100	3.43	
256	1.34	Middle C
512	0.67	One octave above middle C
1000	0.34	
3000	0.11 ⎫	Frequencies at which
4000	0.086 ⎭	hearing is most acute
5000	0.069	
10 000	0.034	
20 000	0.017	Highest frequency audible
Ultrasound in water at velocity of 1500 ms⁻¹		
750 000	0.002 ⎫	Frequencies used in
1 000 000	0.0015 ⎬	ultrasound therapy
3 000 000	0.0005 ⎭	

6.2). Some other mammals can hear at slightly higher frequencies, dogs for example. It is well known that bats use ultrasound – frequencies around 50 000 Hz – for echo-location of objects in their path and for hunting insect prey. The higher-frequency ultrasonic beams are more suitable than low-frequency sound for echo-location because they spread out less, that is, they form narrower beams and hence are more precise. This reflection of sonic waves – echoes – is familiar in sound waves and discussed in Chapter 3. The audible squeaks emitted by bats are, obviously, of lower frequency, around 5000 Hz, and are concerned with sexual behaviour rather than echo-location.

Musical notes or tones are distinguished by being sound of different frequencies. Thus middle C is 256 Hz (or 262), the C above 512 Hz (or 524) and the C below 128 Hz (or 131). Similarly, all the musical notes have definite frequencies which have clear mathematical relationships. For scientific purposes a standard scale is used but in music the actual pitch (frequency) of the note has never been considered of as much importance as the ratio of the frequencies of the notes. Thus, historically, the same note, say middle C, may have been played at slightly different pitches. The frequencies found on modern pianos and standardized by an international agreement about 40 years ago are those given in brackets above. The importance of the ratio of the frequency of two notes, called the musical interval, is well understood by musicians. The intervals are given special names. The ratio between the top and bottom notes of a scale, say middle C to the C above, is 2:1 (262 : 524 Hz) and called an octave.

That the same note can be played on different instruments and will sound entirely different is well recognized. This is due to the fact that single frequencies are almost never produced by a musical instrument. The strongest frequency is called the fundamental frequency and determines the pitch but there are many weaker, different frequencies called overtones or harmonics. It is these overtones that give the note its characteristic quality or timbre. The construction of musical instruments affects the strength and frequency of these overtones so that even superficially identical instruments can sound slightly different. Much skill goes into the making of instruments that will generate the most pleasing sounds.

The frequencies employed in most speech and music lie between 100 and 4000 Hz and these are the frequencies to which the ear is most sensitive. The wavelengths of sound waves in air and for ultrasonic waves in water are given in Table 6.2. It will be realized that sound waves spread out very widely in many situations as sound can easily be heard around corners. This occurs because of diffraction. When part of the sound wave front is blocked the waves are able to bend round. In other words, obstacles in the path of the waves are able to alter their direction of propagation. This occurs with all waves, e.g. water waves and electromagnetic radiation. For waves passing through a gap the amount of spreading depends on wavelength. Thus if the wavelength is much less than the gap, very little diffraction will occur. Hence shorter-wavelength ultrasound tends to behave more like a beam of radiation travelling in a straight line. The reason light appears to travel as a rectilinear beam is because it has only a tiny wavelength (see Chapter 8). Very narrow diffraction grating slits are needed to show that it also can be diffracted. Loudspeakers used in large halls are often designed to allow the most important sound frequencies to spread horizontally. Tall, narrow speakers allow the sound to spread out since their width is made less than the wavelength of many of the sounds to be heard while their height is greater, preventing too much energy loss upwards over the heads of the audience (Table 6.2).

INTENSITY OF SOUND

It has already been noted that the rate of energy flow of sonic waves can be measured in watts per square metre (W m^{-2}) which is proportional to the frequency times amplitude squared. As the energy of a sound is increased it increases in perceived loudness but there is not a simple linear relationship between the energy the sound has and how it is recognized by the ear–brain system. The nervous system is much more responsive to very quiet sounds than it is to very loud sounds so that measures of subjective loudness are made on a logarithmic scale called the *decibel scale*. This is based on the energy of the faintest sound that can be heard, which is 10^{-12} W m^{-2}. The scale is from 0 decibels (dB) to 120 dB, which has an energy of 1 W m^{-2}. This is the point at which the sound is so loud that it is described as pain. Thus the ratio of the faintest detectable sound, the threshold of audibility, to the maximum tolerable sound energy is a trillion (a million million or 10^{12}). It is not recognized in this way by the ear. As might be expected, hearing ability varies somewhat with frequency, being most acute around 3000–4000 Hz and wavelengths around 10 cm (Table 6.2). Hearing acuity tends to decrease with age and sometimes old age or disease may limit hearing to particular frequencies. For examples of the sound level of familiar sounds, see Table 6.3.

Table 6.3 Sound levels

	Sound in dB	Intensity in W m^{-2}
Threshold of audibility	0	10^{-12}
Whispering at 1 m	20	10^{-10}
Home	40	10^{-8}
Normal conversation	60	10^{-6}
In car in traffic	80	10^{-4}
Machine shop	100	10^{-2}
Pneumatic drill	120	1.0

THE VOICE

The ability to produce speech by controlling breathing, vocal cords and speech cavities is learned in early childhood. It is an ability unique to humans and competence to learn and use language appears to be innate. To produce speech, it is necessary both to produce sound and to control it to form definite recognizable phonemes. A phoneme is the smallest unit of sound that forms a functional part of language. All languages are built up of their own set of phonemes. English has some 38 which only roughly correspond to the alphabet. Several letters are used with more than one sound, for instance, the 'a' in father, in hat and in fall.

Airflow from the lungs passes over the nearly closed vocal cords, causing them to vibrate, producing sound at many frequencies. The main one is about 125 Hz in men and rather higher in women because they have smaller vocal cords and a smaller larynx. This voiced sound is modified in the pharynx, oral and nasal cavities by altering the cavity shape to cause particular harmonics to resonate and others to be suppressed. This is effected by moving the lips, soft palate and particularly the tongue.

The importance of the control of the voice cavities can be realized when it is noted that tensing the vocal cords produces high-pitched, falsetto speech but leaves the phonemes and hence speech normally comprehensible. Similarly, after excision of the larynx, near-normal speech is possible by using the eructation of swallowed air to provide voiced sound. On the other hand, damage to the oral or nasal cavity, palate or tongue can cause complete loss of comprehensible speech.

PERCEPTION OF SOUNDS

It is not merely the physical characteristics of sound waves that determine what is perceived by the brain. This is evident when comparing the experience of listening to a speaker in a noisy room with a tape recording of the same event. During the live event the extraneous sounds are automatically ignored. Similarly the higher centres are able to dictate which sounds should be attended to, enabling listeners to select what speech or music they wish to hear from a mass of other sounds of equal or greater intensity.

Possession of two ears allows some degree of sound location. The difference between the times of arrival of a sound at each ear is used by the brain to indicate the direction of the sound. Since the distance between the ears is something less than 15 cm, the brain must be able to detect a difference of, at least, rather less than half a millisecond. In fact it is suggested that differences as small as 30 μs can be recognized (Freeman, 1968). Further evidence of the sound source is gained by comparing the intensity as the head is moved to different positions.

Since the pitch of a note depends on its frequency, which is the number of waves reaching the ear in unit time, it would be expected that the note would change if the source and listener were moving relative to one another. This is often experienced and is known as the *Doppler effect*. It occurs in all wave systems. For sound waves in air the effect often occurs when a horn is sounded continuously from a speeding car passing a stationary (often petrified!) pedestrian. As the sound approaches, the wavefronts appear to be closer together because the source has moved closer, between each wave. Similarly, as the source recedes, the waves are further apart. Thus the approaching sound has a higher frequency, hence pitch, and the receding sound a lower frequency.

Beats

When two sounds of nearly, but not exactly, the same frequency having similar loudness are played, a beating or rising and falling intensity can be heard. This effect is due to the fact that where the two sound waves are both in compression they will cause greater compression but where one is compressed and the other rarefied they will cancel out, leading to diminished sound. This effect will occur with any wave system and is described as interferential current in Chapter 5. As noted, the beat frequency is the difference between the two individual frequencies. This effect is sometimes used to tune piano strings to the note of a tuning fork.

PRODUCTION OF SONIC WAVES

The ways in which sounds are produced are very familiar and involve some form of mechanical vibration. The vibrating ruler, considered in Chapter 3, provided an example of regular vibratory motion which, as has been considered above, gives a constant pitch or musical note. It is the regularity that distinguishes musical notes from noise. Most musical instruments produce notes by vibrating either strings or columns of air. The strings may be plucked, struck or bowed, causing them to vibrate and produce standing waves (see Chapter 3) – and their overtones – at frequencies which depend on the distance between the two fixed ends of the string as well as the material and diameter of the string. The columns of air are caused to vibrate in either open or closed tubes by a disturbance started at one end by eddies, caused when a stream of air is blown at a sharp edge or a reed is allowed to vibrate. The column of air in the organ pipe or other wind instrument can be made to vibrate in different ways. One way is to alter the relative length of the tube by closing and opening holes with the fingers.

To produce ultrasound waves, simple mechanical methods are unsuitable, although special whistles or sirens can be used for frequencies up to about 80 kHz. Beyond this, devices that have higher frequencies of natural vibration are needed. The only way to provide continuous energy at such high frequencies is by means of electrical oscillations. In certain materials mechanical stresses cause electrical polarization. Opposite electrical charges accumulate on the outer surfaces of the material. Bone exhibits this property. This was first described in 1880 by Pierre and Paul-Jacques Curie and is called the *piezoelectric effect*. A reverse piezoelectric effect occurs if these substances are exposed to an alternating current when they will change shape at the frequency of the alternating current.

To produce higher frequencies such as those customarily employed in physiotherapy (0.75, 1 and 3 MHz), piezoelectric transducers are used.

These consist of a suitably-sized disc of crystalline material that will alter in shape when an electric potential is applied across it. If the a.c. frequency exactly matches the natural resonant frequency of the crystal it will oscillate, getting thicker and thinner in time with the alternations of current. Although the effect occurs with many types of crystal (natural quartz was used originally), synthetic ceramic crystals made from mixtures of salts such as barium titanate and lead zirconate titanate (PZT) are utilized these days. The polarity of these materials as well as their precise shape can be controlled during their manufacture. Synthetic crystals only require a small voltage to achieve these effects so a step-up transformer is not required.

If the transducer crystal is in air, very little energy is transmitted, for reasons which are explained below, but if one face of the device is bonded to a metal surface which is in contact with a liquid or the tissues then mechanical wave energy is transmitted.

The word transducer refers to any device that changes one form of energy into another but is often specifically applied to wave energy transformations. In the case described, electrical energy is being converted to mechanical energy. Of course, similar devices can operate in the opposite way. Thus an ultrasound probe converts sonic vibrations into electrical signals which can be displayed as a measure of energy. Similarly, a continuous compression force can be measured by a piezoelectric crystal acting as a force transducer in a force plate or electric dynamometer.

ULTRASONIC GENERATORS FOR PHYSIOTHERAPY

A piezoelectric crystal is fitted into a metal casing which forms the treatment head of the ultrasound apparatus. The actual PZT disc is often about 3–4 mm thick and is bonded to a metal plate. On older units each treatment head was constructed to have a desired natural frequency of oscillation, say 1 MHz. On more modern machines one treatment head is capable of supplying different frequencies of ultrasound. This is achieved by driving the crystal at both its fundamental frequency and a harmonic. It is important to remember that the *effective radiating area* (ERA) of the treatment head is always smaller than its geometric area. The casing is made waterproof and incorporates a handle, the whole being connected by a suitable length of wire to the apparatus.

The basis of the driving circuit is the oscillator which produces sinusoidal alternating current of the appropriate frequency in a way that has been described in Chapter 4. As noted, quartz crystals require higher voltages than synthetic ones and, coupled with the fact that PZT crystals are more robust, the latter have become the preferred choice.

The output of therapeutic ultrasound sources is often applied in pulses and an astable multivibrator circuit is provided to control the output by turning the output from the oscillator on and off to give pulses of the required duration. Many modern generators are based on integrated circuits (see Chapter 4). A block diagram to illustrate these points is shown in Figure 6.2. Switches to turn the apparatus on and off and to change from continuous to pulsed mode are added. Most important is a variable resistance to control the power applied to the crystal and hence the excursion of the crystal and metal plate of the treatment head. This is called the intensity control and dictates the amplitude of the ultrasonic wave. This intensity is measured in watts per square centimetre (W cm^{-2}). Most modern machines can display, on a meter or digitally, both power and intensity, but only one of these at a time. This measurement is the output applied to the crystal and not that of the crystal itself. Thus, if the crystal has failed or is defective this is not always evident from the output display. On investigation it was found that some therapeutic ultrasound units in clinical use did not conform to their stated output (Hekkenberg *et al.*, 1986; Docker, 1987; Lloyd and Evans, 1988).

Timing arrangements are also provided to set a precise length of treatment. Separate indicator lights are often provided to show that the mains circuit and treatment output are on. A useful addition on some machines is a sensor which indicates inadequate contact between the treatment head and the tissues. This is achieved by constantly monitoring the electrical output to the transducer. If there are changes, which will occur if the impedance alters because of inadequate contact, the output is immediately reduced by changing to a pulsed mode with a short (say, 10%) duty cycle and this is indicated on the machine. Usually the timer stops also and recommences automatically when proper

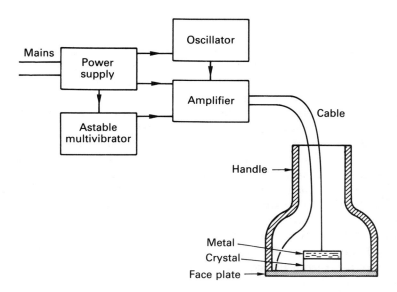

Fig. 6.2 Block diagram of ultrasonic apparatus for therapy.

contact is restored. This is recognized because the crystal loading is being sampled at each pulse.

Therapeutic ultrasound generators nearly all provide pulsing which can be varied. Many give 2 ms pulses and allow the pulse interval to be varied. This is described either as the mark : space ratio – the ratio of pulse length to pulse interval – or as the duty cycle, which is the pulse length as a percentage of the total period (Table 6.4). Some consider 16 Hz to be a fundamental frequency of the cellular calcium system and as such this frequency and its multiples may have particular therapeutic significance.

Table 6.4 Some examples of pulsing

Pulse length (ms)	Pulse interval (ms)	Mark : space ratio	Duty cycle (%)	Pulse frequency (Hz)
2	2	1 : 1	50	250
2	6	1 : 3	25	125
2	8	1 : 4	20	100
4.17	16.67	1 : 4	20	48
12.5	50	1 : 4	20	16

Pattern of the output from therapeutic ultrasound sources

It was noted earlier that the way sound spreads out depends on the size of the sound generator in relation to the wavelength (see Table 6.2). This shows that the wavelength of sound in air is much larger than the voice cavities. From the same table it is evident that the wavelength of therapeutic ultrasound in water or tissues is much smaller, e.g. 1.5 mm, than the treatment head diameter of, say, 4 cm. This means that the sonic waves are projected as a more or less narrow beam. However, this beam is far from regular. The near field or Fresnel zone near to the treatment head is an area in which the intensity of the sonic beam is particularly non-uniform. The far field or Fraunhofer zone, on the other hand, is a region in which the ultrasonic energy tends to spread out so that the intensity falls with distance but is relatively regular.

The reasons for this non-uniformity of the near field are that the transducer face does not move uniformly and since it is perhaps 20 or 30 wavelengths in diameter, different parts of the face will contribute waves slightly out of step, mixing with the other waves after slightly different times. This results in a pattern of high and low intensities (Fig. 6.3) which is highly variable. The minima and maxima can be calculated (Frizzell and Dunn, 1982; Ward, 1986). The extent of this near field can be worked out from the square of the radius (r) of the transducer face and the wavelength.

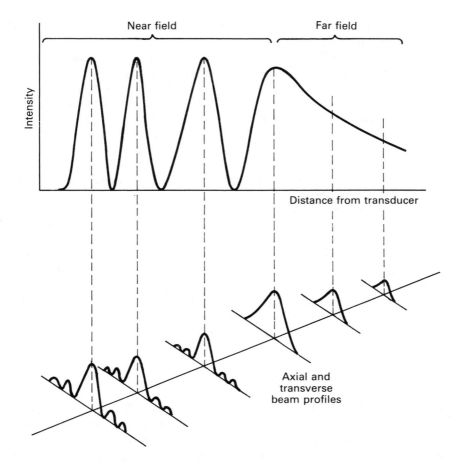

Fig. 6.3 Axial and transverse beam profiles.

Thus:

$$\text{Depth of near field} = \frac{r^2}{\lambda}$$

A 4 cm diameter transducer at 1 MHz in water or tissues would have a near field of 26.6 cm ($r = 20$ mm, $r^2 = 400$ mm, wavelength $= 1.5$ mm). At 3 MHz with a shorter wavelength the near field would be still longer. Thus, from a therapeutic point of view, all effective treatment is delivered by energy in the Fresnel zone. While this zone is approximately cylindrical the distribution of energy transversely across the field is also far from uniform. It has been found that some 70% of the energy is confined to an area about half the diameter of the transducer (Fig. 6.3). This non-uniformity is expressed as the beam non-uniformity ratio (BNR). It represents the ratio of the peak intensity to the average intensity. The lower the BNR, the more uniform the field, with a theoretical minimum of 4.

These characteristics of the ultrasonic beam make it essential that the treatment head be moved in a regular manner during the application of ultrasound therapy if an even distribution of energy is to be achieved and 'hot spots' avoided. What has been noted above refers to ultrasonic energy travelling in a homogeneous medium with negligible absorption. In real treatments, the additional irregularities and absorption might serve to diminish the peaks and troughs somewhat.

The reason why the far field becomes more regular is that, at these distances, variations between different parts of the transducer face and this point become negligible.

These characteristics of the ultrasound beam show the need for a more sophisticated way of describing the output. Thus, as already noted, the intensity is usually given in watts per square cm, i.e. the energy crossing unit area in unit time. To be useful, the area should be specified – usually the effective radiating area – and then it should be called the space-averaged intensity. Further, the average intensity over time may be quoted; either the time-averaged point intensity, which is the average intensity at a point over a given time, or the space-averaged time-averaged intensity, for which both area and time should be specified.

TRANSMISSION OF ULTRASOUND

It has already been noted that sonic waves can be reflected. The reasons why wave energy is reflected when it meets a new material were discussed in Chapter 3. It was seen that waves were reflected at an interface if there was a difference in acoustic impedance of the two media. In the case of sonic waves the impedance (Z) of any medium depends on the density and elasticity of the medium and can be found as the product of the density of the material and the velocity of sonic waves in the material. Table 6.5 gives examples of density and impedance of various substances. What matters in respect to reflection is the *difference* in acoustic impedance. The enormous difference between air and steel can be seen in Table 6.5; one would expect massive reflection to occur at an interface between these materials. (Other metals would be similar to steel.) Thus if the transducer plate of an ultrasound source is in air it can be calculated that some 99.99% of the energy is reflected back into the transducer. Because of this, it is necessary to use a suitable couplant to transmit ultrasonic energy from the treatment head to the tissues. The impedances of all the soft tissues, muscle, fat, blood etc., are very close to one another and close to that of water. The exception shown in Table 6.5 is bone. It is thought that some 25% of ultrasound energy is reflected at the bone–muscle interface (Table 6.6). Further, because bone is a solid, some transverse sonic waves can occur. This is

Table 6.5 Acoustic impedance of some substances

Substance	Density (kg m^{-3})	Velocity (m s^{-1})	Impedance (kg m^{-2} s^{-1})
Air	0.625	343	213
Fat	940	1450	1.4×10^6
Water	1000	1500	1.5×10^6
Muscle	1100	1550	1.7×10^6
Bone	1800	2800	5.1×10^6
Steel	8000	5850	47.0×10^6

Figures taken from Ward (1986).

Table 6.6 Percentage reflection at various interfaces

Interface	% reflection
Water–soft tissues	0.2%
Soft tissue–bone	15–40%
Water–glass	63.2%
Soft tissue–air	99.9%
PZT–air	99.9%

PZT = Lead zirconate titanate.
Figures taken from Williams (1987).

believed to contribute to local heating which is thought to be the cause of the periosteal pain occasionally induced by ultrasonic therapy. Table 6.6 shows examples of reflection at some interfaces.

7.   *Thermal energy*

The distinction between hot and cold is universally understood and the idea of a thermal energy gradient is widely recognized but the fundamental nature of heat energy is often poorly comprehended. This is at least partly due to confusion about the use of words. When an object is heated it is given energy so that its internal energy increases. That is to say, there is more energy in the microstructure of the material.

To understand the meaning of internal energy and the processes of heating it is necessary to consider the nature of the microstructure of matter. This was discussed in Chapter 2 and reference should be made thereto, particularly to Figure 2.5 which is pertinent to much of the following discussion. In summary: matter is found in three states – solids, liquids and gases. Solids are formed of densely and regularly packed collections of atoms or molecules allowing only small movements of the constituents. The motion allowed in liquids is rather greater but in gases the wide spacing of atoms allows quite large random movement to occur.

Thus in solids there is a vibration as molecules and atoms move together and apart. An analogy, described on page 21, Chapter 2, was used to illustrate this concept in which the atoms or molecules were considered as rubber balls containing magnets. These were, therefore, constantly attracted to one another but repulsion occurred as the rubber was distorted. One can thus imagine the atoms oscillating – bouncing to and fro – like the pendulum discussed in Chapter 3. There is a constant exchange between potential energy – as the rubber is compressed or having no effect at the other extreme – and kinetic energy – as the molecules are moving apart or together. This concept, illustrated in Figure 2.5, is central to the explanation of internal energy.

An increase in internal energy is an increase in this oscillatory motion. The molecules gain both kinetic energy and potential energy in that they move faster and further. They are, of course, moving both further apart and closer together. A decrease in energy leads to the molecules moving less.

It is this internal kinetic energy which is recognized as heat; a body with more internal kinetic energy has a higher temperature. The idea of temperature is well recognized as the degree of heating of a body – the hotness of the body. It will also be evident that the total quantity of energy will depend on both the degree of heating and the amount of matter involved. The internal energy of matter can be altered both by heating, for example by putting a body in contact with another hotter body, or by doing work, say by hammering the material vigorously.

TEMPERATURE

As noted already, the level or degree of heat is measured as temperature. Humans judge temperature with receptors in the skin and the perceptions are largely relative, not absolute. Thus quite small differences in the temperature of two surfaces can be recognized if they are compared at the same time. It is a common experience that tepid water feels hot to the hand that has just been removed from cold water but cold to the hand just out of hot water. The recognition of contrasts is a general feature of perception in the nervous system. Compare the recognition of musical notes described in Chapter 6.

Mercury-in-glass thermometers are familiar and work on the principle that heat causes expansion of matter (see later) and that mercury expands more than glass. Thus, if the mercury is caused to expand along a very narrow glass tube, quite small temperature increases will cause a visible movement of mercury along the tube. Temperature measurements are made by taking two fixed points, one being the boiling point of water and the other the melting point of ice, and dividing them into 100 equal parts called degrees. Why these are fixed points is considered later. This scale was originated by Andreas Celsius (1701–1744), a Swedish astronomer, in 1742, and is therefore known as the Celsius scale, abbreviated to C. Formerly this was known as the centigrade scale in English-speaking countries, centigrade meaning divided into 100 units, but Celsius has been adopted by international agreement. The mercury-in-glass thermometer was invented around 1714 by Gabriel Fahrenheit (1686–1736) who used the freezing point of brine as his zero and normal human body temperature as his other fixed point, dividing the interval into 96 degrees. This resulted in the freezing point of water being 32°F, its boiling point 212°F and the subsequently corrected body temperature being 98.4°F. The Fahrenheit scale is no longer used for scientific purposes but is still found on some domestic equipment.

It has been explained that temperature increases with the energy of atoms and molecules so that there is no theoretical upper limit. There is, however, a lower limit since a lower temperature is associated with less motion so that ultimately there is a point at which no motion occurs. This is called *absolute zero* and is −273.15°C. In fact, a very small amount of kinetic energy remains which cannot be further reduced so that absolute zero is the lowest temperature that can exist. The logical kelvin

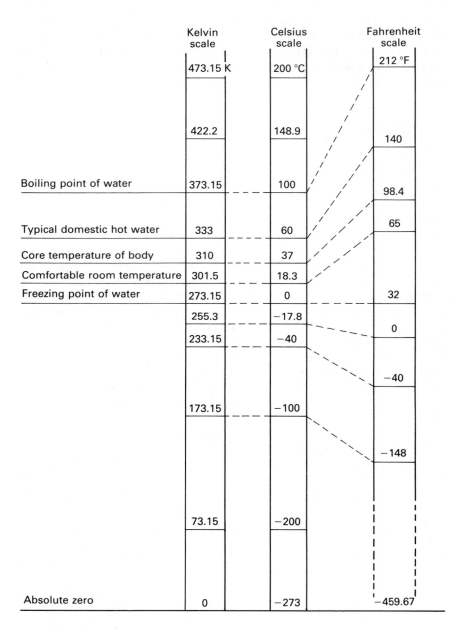

Fig. 7.1 Temperature scales.

scale is the SI unit of temperature, with the same degrees as the Celsius scale but absolute zero being 0 K. The freezing point of water therefore becomes 273.15 K. The scale is named after Lord Kelvin (1824–1907), formerly William Thomson, who deduced the effect. These scales and their relationships are illustrated in Figure 7.1, together with some important temperature points. Conversion of Fahrenheit to Celsius temperatures and vice versa is easily achieved when it is realized that each degree Celsius is 5/9 of a degree Fahrenheit and the scales are out of step by 32° (see Fig. 7.1). Thus:

$$C = \frac{5}{9} \, F - 32 \quad \text{or} \quad F = \frac{9}{5} \, C + 32$$

QUANTITY OF HEAT

Substances at a higher temperature contain more internal kinetic energy than those at a lower temperature. However, the total internal energy must also be related to the mass of material. This is familiarly illustrated by pouring tea or coffee made with boiling water into a little milk in a cup. The resulting liquid is much cooler at once because the energy is now shared with the cooler milk and cup and over a few minutes is lost to the surrounding air, so that the tea or coffee may be comfortably drunk at around 40 or 50°C. (The controversial culinary question of whether the milk should be added before or after pouring is a matter beyond the scope of this book!) If a bath full of cold water has a cupful, or for that matter a whole kettleful, of boiling water added to it, no temperature rise can be detected. Similarly, a bath of hot water, at say 45°C, will remain warm enough to bathe in for perhaps an hour. These examples show that the amount of internal energy depends both on temperature and on the quantity of matter. Basically temperature is a measure of the average kinetic energy of individual molecules whereas heat quantity is a measure of the total energy, both potential and kinetic, in the total mass of matter involved.

Since energy is being described, the appropriate SI units are joules. As might be expected, it takes different amounts of energy to raise the temperature 1°C in different materials. Thus to describe the heat capacity of different materials it is necessary to specify the energy needed to raise the temperature 1°C for a unit mass of the material. Thus, for water it takes 4.18 kJ to raise 1 kg 1°C. (To be strictly accurate, it is raising the temperature of water from 14.5 to 15.5°C because there is a slight difference depending on temperature.) This is known as the *specific heat* or *specific heat capacity*. The idea of specific heat was introduced by the Scottish chemist Joseph Black in 1760 (Table 7.1). Water has a very high specific heat compared with other common

Physical Principles Explained

Table 7.1 Specific heat

	Specific heat (kJ kg^{-1} °C^{-1})
Water	4.185
Air	1.01
Aluminium	0.904
Copper	0.402
Mercury	0.14
Glass	0.77
Paraffin wax	about 2.7
Rubber	2.01
Whole human body	3.56
Skin	3.77
Fat	2.3
Muscle	3.75
Bone	1.59
Whole blood	3.64

Data from Sekins and Emery (1982).

substances, which means that it needs a good deal of energy to raise water temperature and also that hot water will store much heat per unit mass. This fact has considerable climatic and economic significance. The sea temperature changes much more slowly than that of the land so that coastal areas and islands tend to have warmer winters and cooler summers than places of similar latitude in the middle of large continents. On a smaller scale, water is often used as a heat buffer or heat store. For example, a hot water bottle containing, say, 1 kg (1 litre) of water at 90°C would give 209.25 kJ of energy in falling to 40°C (4.185 kJ × 1 × 50). Compare this with the same quantity of mercury which would lead to a release of just 7 kJ over the same temperature range. This, aside from the fact that mercury is highly toxic and wildly expensive, makes it unsuitable for filling hot water bottles!

Formerly a unit called the calorie was used to describe the amount of heat needed to raise the temperature of 1 g of water 1°C so that 1 kcal is the equivalent of 4.18 kJ. (In dietetics a unit called the Calorie, with a capital C, has been rather confusingly used to denote 1 kcal or 4.18 kJ.)

The considerable range of specific heat capacities can be seen in Table 7.1. Adding the same amount of internal kinetic energy to each – i.e. transferring the same amount of heat to each – will lead to different temperatures being attained. These differences are partly due to the number of atoms. A kilogram of aluminium will contain more atoms than a kilogram of copper because each aluminium atom is lighter.

Similarly, mercury atoms would be heavier still so that fewer would be present in a 1 kg block. It will be seen from Table 7.1 that the specific heat capacity of aluminium is rather more than twice that of copper and about six times that of mercury. The substances with more atoms take more energy to set them in motion which partly, but not wholly, accounts for the different specific heat capacities.

Specific heat capacity increases directly, but not regularly, with increasing temperature. In general, the specific heat capacity increases markedly at the lower temperatures but less at higher. The temperature at which this change occurs varies in different elements and also contributes to the specific heat differences.

Furthermore, there are differences in the specific heat capacity which depend on whether it is measured at constant pressure or constant volume. For solids and liquids it is always measured at constant pressure which differs very little (2 or 3%) from the value at constant volume. The value at constant pressure is always slightly greater than that at constant volume because when the material expands (see later) it does work in pressing back the surrounding air and this energy has to be provided above and beyond the energy needed in raising the temperature. In the case of gases the difference is much greater, so that for air at room temperature the specific heat capacity measured at constant pressure is about 40% greater than that measured at constant volume. Since the human body is about 70% water, it would be expected to have a specific heat capacity close to that of water. As can be seen from Table 7.1, it is about 3.56 kJ kg^{-1} °C^{-1} (Sekins and Emery, 1982). This means that it takes quite large quantities of energy to alter the body temperature significantly. As will be seen later (page 171), there are complex physiological mechanisms to maintain a constant body temperature. This process is made easier due to the relatively large heat capacity of the adult human. For example, it would take some 178 kJ of additional energy to raise the body temperature of a 50 kg woman just 1°C. Notice that this would be additional to the energy being continuously generated in the body. In fact, even at rest the basal metabolic rate leads to the emission of heat.

EXPANSION

Apart from causing an increase in temperature, adding energy to the microstructure of matter, i.e. heating, can cause other changes. The most obvious is an increase in volume which is expansion in all dimensions. There are many homely examples of expansion due to heating – the mercury in a clinical thermometer, already noted, is an obvious one. Experiment would show that the line of mercury increases in length as a function of the increase in temperature. In fact, for all materials the linear expansivity can be described as the increase in length per unit length per degree rise in temperature, i.e. in metres per metre per degree Celsius. Note that the unit of length does not matter. It would be the same in centimetres per centimetre or inches per inch, so that expansivity is a ratio or coefficient, often expressed simply as per K or per °C (see Table 7.4). To express the volume change a similar method is used: the cubic expansion is the change in volume per unit volume per 1 °C.

The expansion of liquids is approximately 10 times that of solids and in both it increases with temperature in very much the same way as occurred with specific heat. Gases have still higher expansivities.

The reason why expansion occurs with increasing temperature can be seen by considering Figure 2.5. It is usual to suggest that increased kinetic energy leads to the constituent molecules moving further apart. As indicated previously, though, the increased motion is oscillatory and if the energy–separation curve were uniform the average separation of the molecules would not change. However, it is not uniform, as can be seen, so that with increased to-and-fro motion the mean distance between molecules increases. This increased separation occurs between all the molecules, causing an increased volume of the material, hence expansion.

It will also be evident from Figure 2.5 that the midpoint of the oscillation will not only change but will change along a curve, so that the average separation will be a little greater at higher temperatures. This accounts for the slightly greater expansion at higher temperatures.

Causing expansion by heating is the major means of converting heat energy to mechanical energy, particularly heating gases. Many familiar engines work in this way, e.g. the internal combustion engine of cars, steam turbines of power stations and the jet engines of aircraft. A good deal of ingenuity goes into designing engines that convert as much energy into expansion with as little rise in temperature as possible. However, all such energy conversions involve additional heating of the microstructure so that it can never be 100% efficient (see Fig. 4.38).

LATENT HEAT

Another change that can occur when heat energy is added to matter is a change in state. The change from solid ice to liquid water when heating occurs at 0°C is well known and this change occurs without any increase in temperature. Similarly, water at 100°C can be converted to steam (a gas) at the same temperature by further considerable heating. The same conversion occurs in other substances but involving different quantities of energy (see Table 7.2 and Fig. 7.3). The reverse also occurs, so that condensation of steam to water gives out heat energy at 100°C, as does the formation of ice from water. The discovery that ice melted due to heating, without raising its temperature, was made by the same Scottish chemist, Joseph Black (1728–1799). He called the heat that apparently disappeared into the material *latent heat*.

Table 7.2 Specific latent heat of fusion and vaporization for some substances

	Specific latent heat of fusion (kJ kg^{-1})	*Specific latent heat of vaporization (kJ kg^{-1})*
Carbon dioxide	189	932
Mercury	11	296
Tungsten	192	4350
Copper	205	4790
Water	333	2260

As described, the difference between a solid and a liquid depends on the way the molecules are held together – very firmly in the case of a solid, less firmly in the case of a liquid. In order to change a solid to a liquid, i.e. to melt it, it is necessary to break or release the forces holding the molecules together and this takes energy. Of course there are no mechanical links between the molecules; it is the electrical interatomic forces that are counteracted by the added energy. In crystalline solids the molecules or atoms are arranged in regular arrays, as described in Chapter 2, but when some of the bonding forces are reduced by heating, groups of molecules are able to move more freely, tumbling higgledy-piggledy over and around one another. Some molecules are able to leave one group for another. This is a liquid. Further heating, once melting has occurred, leads to a temperature rise but no further change in the average number of intermolecular bonds. At the vaporization temperature, boiling point, all the remaining bonds are broken to allow wide separation of the molecules, forming a gas. Again the heat energy is used in counteracting the intermolecular or interatomic forces so that there is no rise in temperature while it occurs.

To give a simple analogy of what occurs when energy is added and induces a change of state, consider a large glass jar of coffee beans. At rest, the beans occupy the bottom third of the jar and have some regularity in that the rounded, thicker parts of the beans tend to fit into the spaces where the ends of three or four beans touch in the layer below. Thus the beans are, to some degree, in a regular array like a crystalline solid. If the jar is tipped through a few degrees very slowly the beans stay in position and no movement occurs. If, however, the jar is vibrated or gently tapped, the beans slide over one another and if the jar is tilted at an angle the surface of the moving beans becomes horizontal just like a fluid level. With the added energy of vibration the beans are now behaving like a liquid. If the jar is now shaken very vigorously (with the lid on tight!) beans fly about randomly occupying the whole volume of the jar. This is like a gas with widely separated, rapidly moving molecules driven by the additional energy. Notice that it is both the kinetic and potential energies of the molecules that make up internal energy. The kinetic energy is measured by temperature but it is the potential energy which is altered in melting and freezing or in boiling and condensing so that there is no temperature change. The word latent in its sense of 'remaining hidden but still present' is almost synonymous with 'potential' and is thus highly appropriate – well chosen by Joseph Black. Since melting involves only some of the bonding being disrupted and vaporization involves the loss of all bonding it might be expected that more energy would be needed for the latter process. This is always the case and is shown in Table 7.2. The energy needed to effect these changes in some substances, in kilojoules per kilogram, is shown in the table and called the *specific latent heat of fusion (melting) or vaporization*. The specific latent heats of fusion and vaporization are often given in calories per gram, being, respectively, approximately 79 and 539 cals g^{-1} for water. The large amount of energy stored as vaporization can be seen from the table and explains why so much power can be derived from steam turbines.

There is another interesting and far-reaching characteristic of water. Liquids have rather more space between molecules than solids but much less than in gases. Almost all liquid states are therefore less dense than the solid states. However, water is exceptional in that the molecules are more closely packed together at around 4°C than when the crystalline solid, ice, is formed. Thus ice will float on water.

EVAPORATION

It must be understood that the molecules of substances do not all have the same kinetic energy at any given temperature. It is the average kinetic energy that determines the temperature but this average is compounded of millions of molecules with a range of different temperatures. This is illustrated in Figure 7.2. Note that for the two temperatures illustrated, not only do the average energies differ but also the distribution changes, as shown. In this, the number of molecules as a fraction of the total number is plotted against the kinetic energy. Even with a solid, such as ice, a very few molecules will have kinetic energies well above the average. A few molecules at the surface will have sufficient kinetic energy to escape the bonding forces, the potential energy, and form some water vapour. Of course, some water vapour molecules present in the surrounding air will be trapped on the surface of the ice so that a dynamic equilibrium is reached. This is, of course, a change of state from solid to gas without passing through a liquid state and is known as *sublimation* (Fig. 7.3).

For a liquid, such as water, the temperature is higher and the average kinetic energy of the molecules is therefore greater, leading to many more molecules having enough energy to escape from the surface. Now, clearly, the escaping molecules are those with the highest kinetic energy so that the average kinetic energy of the liquid, its temperature, will fall.

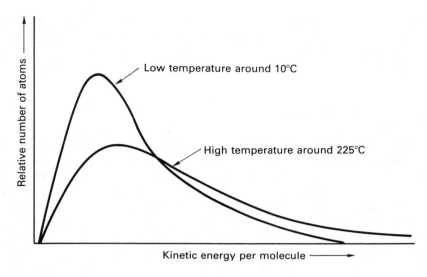

Fig. 7.2 The distribution of kinetic energy for two different temperatures.

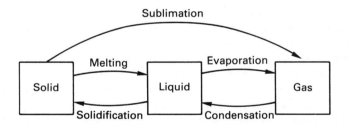

Fig. 7.3 Changes of state.

As long as the evaporated molecules are being swept away, say by a draught of air, the liquid will become cooler and vaporization will diminish. In many real situations the liquid draws energy from the surroundings, its container and the air, so that the system comes into balance. This mechanism, cooling by evaporation, is widely understood not least because it is a major way in which the human body loses heat. In fact it is the only way heat can be lost when the environmental temperature is higher than core body temperature (see page 172). It should be noted that evaporation leads to the liquid losing energy and suffering a fall in temperature because only molecules with high kinetic energies are lost. If those with average kinetic energy were lost there would be an energy loss but no lowering of temperature.

The latent heat of fusion of wax is utilized for therapeutic heating, although it is not large at 35 kJ kg^{-1} (Sekins and Emery, 1982). The melting of ice is used for cryotherapy and from Table 7.2 it can be seen that 333 kJ kg^{-1} is drawn from the tissues. Thus, for example, if 100 g of ice melted in an ice pack that had been applied to an ankle for 5 minutes, the total heating of the ice pack by the skin would be 33.3 kJ at a power of 111 W.

ENERGY CONVERSIONS

Figure 4.38 illustrates some energy conversions, showing that forms of energy can be changed to other forms in various familiar situations. It will be noted that in those situations in which mechanical, chemical or electrical energies are interconverted, some heating also occurs. In other words, such energy conversions cannot be 100% efficient. Likewise the energy of the microstructure of matter can never be wholly converted to some other form of energy; that is, heating cannot be 100% efficient. Because energy is ultimately ending as internal energy of the microstructure of matter – heat – this is often considered to be the basic energy form but it is really a tendency to randomization or disorder. Consider the regular to-and-fro motion of the lower limbs during walking. All the molecules of the lower limb are moving up and down, to and fro, in regular cyclical motion. The appropriate muscles are contracting and relaxing regularly to maintain the walking cycle. All the time some of this regular motion is being converted to heat (friction of the leg moving through air molecules, friction of the foot on the ground, heat due to muscle contraction) which is random, disordered movement of molecules. This concept is expressed in a variety of ways as the second law of thermodynamics. If a system is free to change itself it will tend to go towards a more disordered state. In other words, energy is randomized. The degree of disorder is described by the word *entropy*. The entropy of a substance increases if energy is given to it. That the entropy of a closed system increases with time is another way of expressing the second law. Yet another expression of the same concept is to assert that heat can only transfer from a hotter to a cooler body and not in the opposite direction. Consider a metal spoon used to stir some simmering soup. When removed from the liquid, the spoon is hotter because some of the rapid motion of the molecules in the soup has been transferred to the spoon and hence lost from the soup. Thus randomization has occurred.

A consequence of the second law of thermodynamics is that all energy conversions ultimately lead to an increase in molecular disorder. If the universe is a thermodynamically closed system its entropy gradually increases so that although the total energy in the system is the same, its usefulness diminishes. Eventually a time will come when no energy remains available in a useful form. This state is called the heat death of the universe. (Fortunately, if this stage ever occurs, human beings will be long since gone!)

Having considered the second law of thermodynamics it may be asked, what of the first? This has been implied already and noted in

Chapter 2. It asserts that heat is a form of energy and that energy is conserved when transformations between heat and other forms of energy occur. In a system of constant mass, energy can be neither created nor destroyed. Thus there is a law of conservation of mass – matter can neither be created nor destroyed – and conservation of energy. These were formerly separately developed concepts but in reality mass and energy are different forms of the same thing. The special theory of relativity, postulated by Albert Einstein in 1905, gave expression to this idea. It is usually expressed as

$$E = mc^2$$

in which E is energy, m is mass and c is the velocity of electromagnetic radiation, i.e. approximately 3×10^8 m s^{-1} (see Chapter 8). Thus the conversion of quite small amounts of matter will give enormous quantities of energy, as occurs in nuclear fission. The source of all energy on earth is radiation from the sun which occurs because of continuous conversion of matter to energy. The extent of the energy contained in matter can be glimpsed from the calculation that there is enough energy in 1 kg of matter to operate 1000 one kW electric heaters continuously for 300 years! Looked at from the other aspect, one can say that matter is highly concentrated energy.

ENERGY TRANSFER

Transferring energy through matter or between objects occurs by collisions of neighbouring moving molecules – a process called *conduction*. As noted above, the faster moving molecules or atoms jostle their less energetic neighbours to average out the energy between them. This effect is easily demonstrated, especially in metals in which it occurs with great ease. When one end of a metal bar is heated the other end becomes hot after a little time (painfully familiar to those of us who have left the metal spoon in the simmering soup!). It might be expected that solids with more closely packed molecules would transmit heat energy more readily than liquids or gases and this is broadly true. Metals, however, are exceptionally good conductors of heat (Table 7.3). This is due to the fact that metals contain many free electrons (see page 21) which are given high velocities during heating and are able to move freely in the material. It would not be surprising, therefore, to find that thermal conductivity parallels electrical conductivity and this is true in metals because both mechanisms depend on free electrons. In fact, the ratio of thermal conductivity to electrical conductivity is a constant for all pure metals at a given temperature – the Wiedemann–Franz law. Since thermal energy can be transmitted in non-metals by means other than electron motion, i.e. molecular collisions, and electrical energy cannot, there is no correspondence between them in these materials. Whereas

Table 7.3 Thermal conductivity of some substances

	Thermal conductivity $(W\ m^{-1}\ °C^{-1})$
Metals	
Silver	428
Copper	403
Aluminium	236
Iron	84
Non-metals	
Glass	0.8
Concrete	1.2
Brick	~1
Expanded polystyrene	~0.01
Insulating material	0.04
Wood	0.1
Polythene	~0.4
Water	0.6
Air	0.025
Body tissues	
Fat	0.19–0.45
Muscle	0.54–0.64
Skin	0.33–0.96
Bone	1.16
Blood	0.55

Values are taken from various sources in the section on further reading.

the best electrical insulators have conductivities billions and billions of times less than the worst – i.e. they have enormously higher resistances – the best thermal insulators have conductivities only about a thousandth of the best thermal conductors. The use of the terms conductor and insulator reflects the similarity of the mechanisms of thermal and electrical energy transfer but it should be clearly understood that they are not the same. Whether electrical or thermal conduction is being described should always be made clear if the context does not do so. The factors which determine the rate of heat energy flow by conduction through a piece of material are analogous to those dictating the rate of electric charge flow, the electric current.

Ohm's law will be recalled from Chapter 4, in which the current intensity was determined directly by the e.m.f. (pressure difference across a piece of material) and inversely by the resistance. The resistance, at a given temperature, depends directly on the length of the pathway and the nature of the material and inversely on the cross-sectional area. Similarly, the rate of heat flow depends directly on the temperature difference across the material, the cross-sectional area and the thermal conductivity of the material and inversely on the length of the pathway. (Conductivity is, of course, the reciprocal of resistance; it is the resistance to flow that is used in Ohm's law.) Put simply, thermal conductivity is a measure of the ease with which heat energy is conducted through a given material and differs in different materials (Table 7.3). Since it is a measure of velocity of energy flow it will be in joules per metre per second, hence watts per metre, and since it depends on temperature difference it is expressed per degree Celsius. It can be seen from Table 7.3 that muscle and blood are reasonably close to the value for water, which is hardly surprising, and other highly vascular organs and tissues have similar values. Attention is drawn to the much lower values of subcutaneous fat – around 0.2 W m^{-1} °C^{-1}, which contributes greatly to the thermal control mechanisms of the human body (see page 171). The range of values given for skin in Table 7.3 may seem large but considerable variation would occur depending on the fluid (blood) content of the skin. The outer surface of the epidermis, stratum corneum, seems to have a low, fairly consistent thermal conductivity but the deeper layers and the dermis have a greater value which increases with increasing vascularity (Buetner, 1951).

The markedly greater conductivity of all the metals can also be seen from Table 7.3 – the consequence of having freely mobile electrons, as explained above – and the best electrical conductors, silver and copper, are seen to be the best thermal conductors. It will also be seen that air has a very much lower thermal conductivity than the rest. This explains why almost all practical thermal insulating systems involve maintaining or trapping a layer of air between the surfaces across which heat exchange might occur; examples include cavity brick walls, fibreglass lagging of hot water cylinders or the expanded polystyrene of the body of refrigerators, woollen blankets and jumpers and the fur of the arctic fox! Gases, other than air, also work well as insulators, for example, plastic foams filled with chlorofluorocarbons (CFCs), which are shortly to be banned because of their deleterious effect on ozone in the upper atmosphere. It has been suggested that they might in future be replaced by a new material based on silica aerogels, a silica skeleton with millions of submicroscopic (under 100 nm) air-filled pores. Due to the small size of the air pockets and the low thermal conductivity of the silica skeleton, these have incredibly good thermal insulating properties (Fricke, 1993).

The second way in which heat energy can be transferred is by *radiation*. Adding energy to matter alters the vibrational and rotational levels in the molecules. Some of the energy can be emitted as radiation in the infrared and microwave regions of the spectrum. Greater energy, disturbing the electron configuration of atoms, can lead to the emission of visible radiation. These are all forms of electromagnetic radiation which are described and more fully discussed in Chapter 8. Visible radiation, light, is thoroughly familiar because humans are provided with very sensitive and precise detectors – eyes! However, all bodies, including the human body, are emitting and absorbing a great deal of invisible infrared and microwave radiation all the time. Electromagnetic radiations are oscillations between electric and magnetic properties able to travel in space at a constant velocity of 3×10^8 m s^{-1} in straight lines. The different forms, microwaves, infrared, visible etc., differ only in their wavelength and frequency.

Radiation is, therefore, a way in which thermal energy is transferred. The thermal energy of one body causes electromagnetic energy to travel through space as visible, infrared or microwave radiation which is absorbed by a second body, adding energy to its microstructure, i.e. heating it. The amount and wavelength of the radiation (and hence its kind) will depend on temperature. All bodies radiate and the amount rises proportionally with the fourth power of the Kelvin temperature. Higher temperatures lead to the emission of proportionally more short wavelength radiation. (These relationships are defined in Stefan's law and Wien's law, respectively.) This is familiar in that human cutaneous thermal receptors can detect heating due to the infrared and microwave radiation emitted from heated objects at a distance. Cooling is simply the absence of heat and this is perceived in the skin by separate detectors. Heating becomes more evident as the object gets hotter, say, for example, the heating element of an electric fire. As the element gets hotter still, not only is more heat detected but a proportion of shorter wavelengths start to be emitted, recognized as red light. At higher temperatures orange and yellow light is emitted and further heating leads to the object becoming white-hot (though not domestic electric fires!)

All energy on earth is the result of radiation transmitted to earth from the sun. The sun is continuously converting matter into energy in a form of nuclear fission, giving out enormous amounts of radiation due to its high temperature of around 5700°C (6000 K). The carbon and hydrogen chemical energy bonds stored in oil, gas and coal are the consequence of millions of years of photosynthesis. The current food chains ultimately depend on present-day photosynthesis, as do the climatic cycles.

Thus the two strictly basic ways of transferring heat energy are conduction, in which movement is transferred directly from molecule to molecule, and radiation, in which heat energy is transferred in the form of energy radiation. It is customary to describe the movement of bulk matter as a means of transferring heat. This, of course, can only occur in fluids, i.e. liquids and gases, and is called *convection*. It occurs when the hotter, more rapidly moving atoms or molecules move from one place to another, hence transferring their heat energy with them. There are two ways in which bulk matter might be moved. Firstly, it can be pumped, e.g. the distribution of heat energy around the body as a result of the heart circulating warmed blood or the pumping of hot water through radiators in a central heating system. Secondly, heated liquid or gas becomes less dense than the cooler surrounding fluid and hence tends to float upwards. This is a thoroughly familiar idea in that liquids become evenly heated if the heating is applied from the bottom: smoke rises up a chimney or hot air balloons float. This latter is called thermal convection and the former mechanism is called forced convection.

It will also be realized that evaporation, discussed on page 156, in which the most energetic molecules leave a surface to form a gas, is also a method of heat transfer, a kind of special example of convection.

The transfer of heat energy has been described as a set of separate mechanisms: conduction and radiation only occur where there is a temperature difference, convection only in fluids and evaporation at surfaces. It should be realized that in everyday situations heat exchanges occur in complex chains or networks involving all the mechanisms to different degrees. For instance, the human body tends to lose most of its excess heat at lower temperatures by radiation but if the environmental temperature is close to body temperature than evapora-tion of sweat is the only method (see page 171). Further, consider the familiar central heating radiator in a room. The radiator is heated by pumped hot water (forced convection) which heats the metal body of the radiator which in turn heats the air in contact with it (conduction). This air rises and is replaced by cooler air from the floor (thermal convection), gradually heating the air in the room. The radiator also emits infrared radiations which are absorbed by objects and people in the room (radiation). While both convection and radiation heating will increase with increasing temperature of the radiator, convection will be an almost linear increase while radiation increases exponentially (Fig. 7.4). This means that at low temperatures, convection is much the more

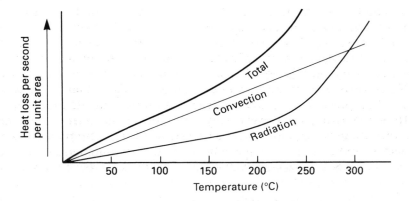

Fig. 7.4 Graph of heat transfer from an object kept at the same temperature above the surroundings, showing the different combinations of convection and radiation.

important process but with increasing temperatures radiation plays a greater and greater part. Domestic radiators have typical surface temperatures around 50–80°C, leading to room temperatures of about 18°C. If a door and window are opened a through draught (forced convection) of cold air will reduce the room temperature.

THERMOMETERS

Fahrenheit's mercury-in-glass thermometer has been mentioned above in connection with the measurement of temperature (see also Fig. 7.1). Any instrument that measures temperature is called a thermometer. By far the most widely encountered mechanism is that of mercury-in-glass (see page 149) which exploits the difference in expansion between the two. Heating the mercury-filled glass bulb causes the mercury to expand, which it is only able to do along the narrow glass capillary tube. The clinical thermometer, invented circa 1818 by T.C. Albutt, a physician, measures temperatures between 30 and 45°C and has a constriction in the glass tube to break the thread of mercury as cooling occurs. This effectively makes it a maximum reading thermometer, allowing accurate reading of the highest temperature recorded. Reading the device is helped by making the cylindrical glass tube act as a lens to enlarge the fine mercury thread. The accuracy of such simple clinical thermometers is surprisingly high and quite sufficient for the measurement of body core temperature. When new, the majority are accurate to within 0.1°C (Cetas, 1982; Togawa, 1985). Mercury-in-glass thermometers do, however, have disadvantages for body temperature measurement under some circumstances. It is difficult to arrange for the continuous monitoring of body temperature whereas other devices (see below) generate an electrical output which is easily connected to a chart recorder. Further, there is always some risk of breakage with the associated dangers of broken glass and mercury contamination. Also these thermometers cannot be made very small and thus have a large heat capacity and slow response.

Alcohol-in-glass thermometers work on the same principle and are often used for measuring room or water temperatures. They are often quite large and the alcohol can be coloured to make them more easily read. Any thermometer that depends on the expansion of liquids will be limited to measuring temperatures between its freezing and boiling points. Mercury remains liquid between −39°C and 357°C so it has a useful range. For low temperatures alcohol thermometers are used because alcohol freezes only at −115°C. In some parts of the world – Siberia, Northern Canada – temperatures below −40°C occur quite regularly. To measure extremely low temperatures, down to −200°C, pentane thermometers are used.

Instead of using the difference in expansion between a liquid and a solid, as in the above thermometers, the difference in expansion between two metals is exploited to provide a robust but not notably accurate thermometer. This is based on a bimetallic strip and is widely used to operate electrical switches to control temperature. Inspection of Table 7.4 will show that there are considerable variations in the linear expansion of a selection of metals. It will be seen that invar, which is an alloy of steel and nickel, has exceptionally low expansivity – about a 20th of brass. The word invar is a contraction of invariate because of this property. It is much used in situations in which thermal expansion would cause problems, such as the balance wheel of watches. If two strips of different metals, such as brass and invar, are riveted firmly together, heating will cause different degrees of expansion in each, causing the bimetallic strip to bend so that the brass is on the outer convex side of the curve. This bending can be made to operate a pair of electrical contacts (Fig. 7.5) to open or close a circuit as the temperature rises or falls. This is extensively used as a thermostat, a device to maintain a constant temperature. To keep an approximately constant water temperature, the heater is controlled by a bimetallic strip switch so that when the water reaches a certain preset temperature the strip bends and opens the switch. As the water cools, so the strip returns to its original position and reconnects the circuit. Such systems are found in numerous situations, e.g. domestic water heaters, electric blankets and heater pads, and the wax baths and hydrocollator heaters found in physiotherapy departments and many other places. This kind of mechanism is called a negative feedback or servo system and is found in many other situations. If the bimetallic strip is made longer and wound into a spiral with one end fixed, the other end can move a pointer over a scale to show temperature, thus forming a thermometer. This conversion of temperature changes to mechanical movement is also used to operate mechanical valves, for example in gas cookers and gas water-

Table 7.4 Linear expansivities

Metals	
Aluminium	$2.6 \times 10^{-5}\ °C^{-1}$
Brass	$1.9 \times 10^{-5}\ °C^{-1}$
Copper	$2\ \ \times 10^{-5}\ °C^{-1}$
Steel	$1\ \ \times 10^{-5}\ °C^{-1}$
Iron	$1.2 \times 10^{-5}\ °C^{-1}$
Platinum	$0.9 \times 10^{-5}\ °C^{-1}$
Invar	$0.1 \times 10^{-5}\ °C^{-1}$
Other materials	
Glass	0.85×10^{-5} per C^{-1}
Silica	0.04×10^{-5} per C^{-1}

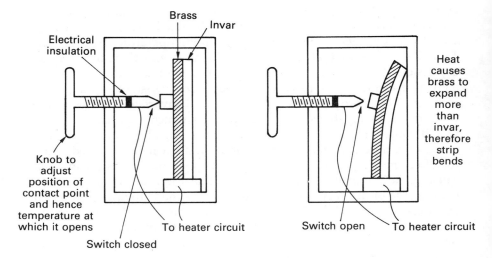

Fig. 7.5 A thermostat.

heating systems and other devices. It is easy to see how adjustments can be made to the point at which the electrical contacts or the valve opens and closes. It is simply a matter of turning a screw to set the physical position at which a certain degree of bending of the strip causes the opening and closing to occur (Fig. 7.5).

As explained on page 153, expansivity is a ratio – length per unit length – so it is not necessary to indicate the units. In Table 7.4 $°C^{-1}$ has been used but it is more correct to use 'K^{-1}'. Note that it is not necessary to refer to '$°K^{-1}$'.

Apart from expansion, other effects that vary with temperature are used to operate thermometers. It has already been noted in Chapter 4 that the ohmic resistance of metals rises with heating. Heating metals causes their constituent atoms to vibrate more extensively in their crystal lattice structures, leading to a larger number of collisions between them and the freely moving electrons. This hampers electron movement, hence the metal has greater resistance. The metal platinum is used in thermometers to make accurate temperature measurements, especially for laboratory work. The electrical resistance of platinum rises by about 0.4% per °C of its value at 0°C and a suitable circuit can provide a very precise measure of resistance and hence of the temperature to be measured. It is, however, somewhat expensive and complex and is mainly used in medicine for experimental work in which the very rapid response and good stability are important.

The most widely used electrical resistance thermometer is the thermistor. This is based on a semiconductor whose resistance decreases exponentially with increasing temperature (see Chapter 4). The electrons in a semiconductor are bound to their atoms at low temperatures but on heating the additional energy allows more and more of them to wander freely from atom to atom, the material thus becoming more like a metal. The decreasing resistance is not linear with increasing temperature – it is exponential – but this can easily be rectified with a suitable circuit. The great advantages of thermistors are the relatively large change in resistance that occurs – about 4% per °C at body temperature (Togawa, 1985) – and their small size. They are mounted in little discs of epoxy resin and can be used as skin thermometers or as thermal probes in hypodermic needles for the evaluation of deep tissue temperatures. They are made of metal oxides or salts such as magnesium oxide. Many modern resistance thermometers provide a digital display for clear reading of their output. Another effect of heating is utilized in the thermocouple thermometer. In these the electromotive force is measured that develops between two different metals if the junction at one end is at a different temperature to the other end. This voltage is proportional to the temperature difference. Various pairs of metals are suitable, such as copper and constantan or iron and constantan (constantan is an alloy of copper and nickel with traces of other metals), or platinum and rhodium. It is necessary to have a known temperature at the reference junction, e.g. melting ice. They can be quite accurate and made very small so that they have a rapid response. A similar device is the p–n junction (see Chapter 4) whose voltage varies with temperature when a constant current is flowing. This gives a much larger voltage per °C (about 2 mV per °C) than a copper–constantan junction and can be made small and is relatively accurate.

Crystal resonators (Chapter 4) also vary in frequency with temperature. Measuring the exact frequency can thus provide an accurate thermometer.

It has been noted above that all bodies emit infrared radiation which is proportional to their temperature. This can be measured in various ways, providing radiation thermometers. The thermocouple described above can be used in this way but photoelectric detectors (Chapter 4) are more usually used. Situations in which very high temperatures are to be measured, such as furnaces, are particularly suited to radiation measurement because the thermometer – often called a pyrometer in these circumstances – can be used at a distance.

Sensing the temperature of quite large areas of the human body surface from a distance can be done by photography of the body with infrared-sensitive film. The film can be made to give a colour response to different temperatures, thus producing a coloured map of the surface temperature. This technique is called thermography and is used to locate and measure the size of hot spots on the surface. These spots may indicate underlying areas of inflammation, in rheumatoid arthritis, for example, or identify areas of greater metabolic activity in attempts at the early detection of tumours. In the same way television video cameras can be made that are sensitive to infrared and used to produce the same sort of picture.

It has been suggested that the microwave radiation emitted by the body can be used to measure temperature in the deeper body tissues. This is possible because microwave radiation penetrates tissues much more effectively than infrared. Although the intensity of microwave radiation given out from the tissue surface is miniscule compared with the infrared – it is about 100 million times less than the peak infrared (Togawa, 1985) – it can be measured with a suitable microwave radiometer.

Liquid crystals, which are forms of regularly aligned molecules that can be realigned under the influence of an electric field, are found in calculators and similar displays. In some kinds a colour change may occur on heating because the long molecules may twist more closely together or move apart. This alteration of spacing alters the wavelengths of visible light that are reflected. These can be used as surface thermometers and have been used in the detection of breast cancer (Davison *et al.*, 1972). They have also been used to make simple thermograms – contact thermography – by putting liquid crystals fixed in a plastic sheet close to the body surface and photographing the display with ordinary instant colour film. This has been tried in connection with the early detection of venous thrombosis.

Another recently developed type of thermometer uses the light carried by optical fibres (see Chapter 8) to transmit the temperature information. There are several types of sensor that can be used: the liquid crystal mentioned above or the decay time of fluorescence of a ruby crystal. Light is transmitted through one optical fibre and the reflected light returned through another fibre. This is sensed and analysed with a suitable electronic circuit to give a temperature reading. These thermometers are mainly for research. Their importance lies in the fact that they are unaffected by electromagnetic fields. Thus they can be used for the direct measurement of tissue temperature during the application of any form of diathermy.

Many other types of thermometer have been developed for special purposes. It is striking that for clinical and domestic use electronic thermometers are not significantly more accurate and certainly no more reliable than the mercury-in-glass type originally developed some 250 years ago! Table 7.5 summarizes some of the different types of thermometers and their principles of working.

Methods of measuring body temperature

By far the most widely used method of assessing the core temperature is the use of a clinical thermometer in the mouth. Kept in the closed mouth for 3 minutes, these give a reading about 0.3–0.4°C lower than the arterial blood temperature. This difference is not clinically important. The main disadvantages of the oral thermometer are cross-infection, which can be easily controlled by the use of disposable covers, and the difficulty of continuous monitoring.

Table 7.5 Summary of thermometers

Type	Change due to heat – principle of working
Mercury-in-glass	Expansion
Alcohol-in-glass	Expansion
Bimetallic strip	Expansion
Platinum wire	Electrical resistance increases
Thermistor	Electrical resistance decreases
Thermocouple	e.m.f. between dissimilar metals
p–n junction	e.m.f. across current-carrying junction
Thermography	Infrared radiation measured
Liquid crystal display	Reflection of visible radiations

The axilla is a convenient site at which to measure core temperature but is usually found to give a lower reading than the oral temperature. To some extent the difference appears to be reduced if the axillary thermometer remains in place for 15–20 minutes. For clinical purposes an axillary thermometer left in place for 10 minutes can read about half a degree lower than the oral temperature.

Rectal temperatures measured with a thermistor or other probe about 10 cm from the anal sphincter are used in hospital. Curiously, the temperature at this site is higher by about 0.3°C than blood temperature in any other part of the body. The changes in rectal temperature tend to lag behind changes in the rest of the body.

Core temperature can also be assessed from the skin by using what is called a zero heat flow method (Togawa, 1985). A thermometer is applied to the skin surface and thermally insulated from a second thermometer and heater. Heat from the skin leaking through the insulation triggers the heater to maintain an identical temperature so that eventually the skin surface reaches core temperature, which is measured.

Skin temperature is commonly measured with a thermistor applied to the surface. These need to be maintained in place to give accurate and repeatable measurements. The use of radiation thermometers has already been noted.

HUMAN BODY TEMPERATURE

Humans have a constant core temperature, i.e. they are homeothermic. The skin and subcutaneous tissue temperatures, especially of the extremities, are much more variable. Human core temperature exhibits a circadian rhythm, being lower in the morning and at its highest in the

Physical Principles Explained

afternoon over a range of about 1°C (Hardy, 1982). Most humans have average core temperatures around 36.8°C but variations between 36 and 38°C are within the normal range. Many children have slightly higher temperatures due to their higher metabolic rate. Factors which cause temporary variations in body temperature are digestion and absorption of food, menstruation and, above all, exercise. Most adults doing moderate exercise experience a body temperature rise of only a degree or so but vigorous exercise can lead to rises up to 40°C for short periods. A prolonged temperature rise is associated with disease and is, in fact, widely used as an objective measure of the severity of the disease. (The German, K.A. Wunderlich, was the first physician in 1858 to introduce the procedure of regular body temperature measurements to chart the course of disease.) This is due to a resetting of the body's thermostat to maintain a higher temperature and higher metabolic rate to increase the effectiveness of the natural defence mechanisms, such as greater activity of the immune system.

The temperature at the tissue surface is usually very much lower than that of the core (Fig. 7.6) and is subject to considerable variation.

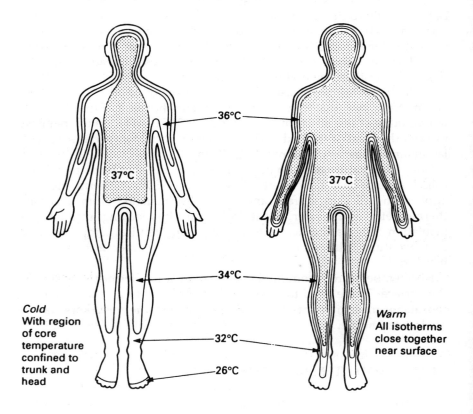

Fig. 7.6 Body isotherms.

Comfortable skin temperatures are around 30°C. Variations of skin surface temperatures are instrumental in controlling the core temperature.

Maintenance of homeothermy

The constancy of body temperature by all homeothermic animals is achieved by the sophisticated control of heat energy loss and gain to maintain a balance. This can be considered like a credit and debit financial balance sheet in which the causes of heat gain are indicated in one column and heat loss in the other (Table 7.6). These effects are not quantified in the table and it is important to realize that some are quite trivial. The major heat gain is from metabolism which is vastly increased during vigorous exercise. For example, the rate at which energy is generated in an average 70 kg man – that is the metabolic rate – is approximately 120 W (Cromer, 1981). This is, of course, an average. During sleep it may be only 75 W but 200–300 W during activity and about 1000 W in strenuous activity. It should also be recognized that some 75% of the energy involved in muscle contraction appears as heat so that any form of muscular activity greatly increases heat production. To maintain homeostasis this additional heat must be lost. At moderate environmental temperature some 50–60% of the heat is lost by radiation. This is infrared radiation of a whole range of wavelengths, with a peak around 10 000 nm (Nightingale, 1959). The human skin surface acts as an almost perfect black body, which means that it absorbs almost all the electromagnetic radiation falling upon it. It also emits a continuous infrared spectrum. The emissivity of a black body is 1, meaning it emits the maximum power per unit area in the wavelength range that that

Table 7.6 Causes of heat gain and loss

Causes of heat gain	*Causes of heat loss*
Basal metabolism	Radiation to the environment
Metabolism of muscle contraction	Conduction to cooler objects
Metabolism of other tissues beyond basal, e.g. digestion	Conduction to air, continually removed by convection
Absorption of radiation from the environment	Evaporation of water from skin – insensible perspiration – vapour carried away by convection
Conduction from hotter objects	Evaporation of sweat – water vapour carried away by convection.
	Exhaled warm air – forced convection
	Excretion of urine, faeces and other fluids

body can emit, at a given temperature. A perfect reflector would have an emissivity of 0. The emissivity of the skin for infrared is found to be 0.97 (Cromer, 1981). It should be noted that visible radiation will be reflected to a greater extent from fair skin than from dark skin, but this has a negligible effect on heat exchange.

At high environmental temperatures heat loss by evaporation of sweat becomes more important and in fact when the outside temperature is the same as or greater than the core temperature, evaporation is the only method of heat loss. This occurs because heat can only be lost from the body surface by radiation or conduction if the body temperature is higher than that of its surroundings. Sweating is therefore essential for survival at these temperatures. If all the sweat is evaporated it provides a very efficient method of heat loss since the evaporation of each gram of sweat takes 2.4 kJ of heat energy. Thus a man producing 1 kg (1 litre) of sweat per hour achieves a cooling rate of some 667 W (2400 J × 1000 g/3600 s = 666.6 W). It is considered that a person acclimatized to a hot climate can sweat as much as 4 litres per hour, giving a cooling rate of some 2.67 kW. In practice it is usual for much of the sweat to be lost from the skin surface before it evaporates so that these high cooling rates are not reached. Evaporation will also depend on humidity (see later).

At moderate temperatures and in the absence of obvious sweating there is a continuous evaporation of water from the skin surface and mucous membrane of the mouth, nasal and pharyngeal passages. This is a further cooling effect. It has been estimated that some 500–700 g of water per day is evaporated in this way, giving about 17 W cooling – about 14% of the total 120 W per day.

Temperature control of the body must really be considered as a two-stage process. The processes just discussed are those involving heat exchange at the skin or mucous membrane surface, which are in themselves uncontrollable. The second stage involves the transfer of body core heat to the body surface, a process which is under physiological control. The difference between the body surface temperature and that of the surroundings is crucial in determining the amount and direction of heat energy flow from one to the other. Thus a skin surface, hotter than the air surrounding it, will lose heat energy by conduction to the air and by radiation. The air in contact with the skin will be moved away by forced or thermal convection, to be replaced by cooler air, to be again warmed by conduction. This process is deliberately inhibited by wearing clothes which effectively trap a layer of air close to the skin, preventing convection, and since thermal conduction through air is very low (see Table 7.3), heat energy loss is diminished. The rate of these processes increases with greater temperature differences (Fig. 7.4), as also does the rate of evaporation. This latter process is also dependent on the amount of water vapour present in the surrounding air. Thus, under conditions of high humidity, sweating is less efficient as a heat loss mechanism – a fact well-recognized by those exercising in humid conditions.

The body surface temperature can, however, be well controlled physiologically so that a large difference can be maintained between the core and the outer 'shell' of the body. This idea is illustrated in Figure 7.6. This can be achieved because of the low thermal conductivity of the tissues, especially fat tissue (Table 7.3). The skin temperature is dependent on the blood flow, in that warmed arterial blood from the core transfers heat energy through the tissues to the skin, i.e. forced convection. This transmits heat energy through the thermal barrier provided by subcutaneous fat. The concept of a temperature gradient between the core and periphery is expressed schematically in Figure 7.6 with the use of isothermal lines, which also illustrate the lower temperatures of the extremities, so that the skin temperature of the hands and feet can be close to environmental temperature.

If the body becomes heated and the core temperature starts to rise, then cutaneous vasodilation causes a rise in skin temperature, allowing greater heat energy loss to rebalance the thermal state. If the body is cooled then vasoconstriction occurs, diminishing heat energy loss. These effects are also illustrated in Figure 7.6. On average women have a thicker layer of subcutaneous fat than men. This ensures that, on the whole, women have lower skin temperatures below about 30°C and higher above (Fig. 7.7). At environmental temperatures above 30°C or so sweating starts to occur (with the body at rest) and this, coupled with the vasodilation, causes the flattening of the skin temperature rise shown in Figure 7.7. In general it seems that sweating starts earlier and is more extensive in men.

Another mechanism that acts to conserve the body core temperature is the countercurrent heat exchange that occurs in the limbs. Warm arterial blood moving from the core to the periphery in medium arteries is able to pass heat energy to the cooler blood returning from the periphery and carried in adjacent veins. The arrangement is evident in the venae comitantes of the limbs. As the arterial blood gradually cools in its passage to the more distal parts it is close to the coldest of the blood returning in the veins. More centrally, the venous blood has been somewhat warmed but it is close to the hottest arterial blood. Thus the temperature gradient between arterial and venous blood remains much the same throughout, facilitating maximum heat energy transfer. It is not simply the effect of arteries and veins being close. Inspection of Figure 7.6 shows that arterial blood would pass through progressively cooler zones of tissue on its journey to the extremities and thus affect the total venous and lymphatic fluids draining centrally. The importance of this arrangement is that it conserves core heat by using it to warm incoming venous blood instead of wastefully heating the peripheral parts. Thus the periphery must remain at a lower temperature. While this arrangement is of undoubted importance in some animals,

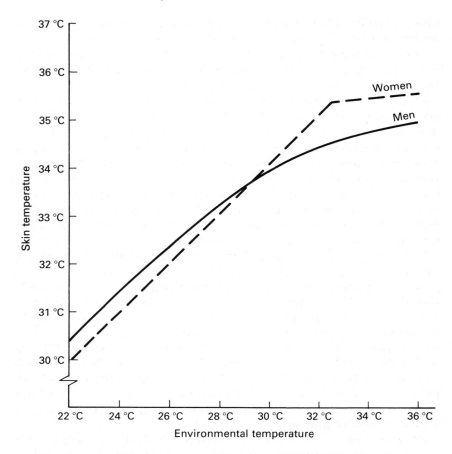

Fig. 7.7 Skin temperatures resulting from environmental temperatures. Modified from Hardy (1982).

especially those in cold habitats, its value to humans is less certain. Estimates of the heat energy conserved vary from a trivial 5% to 50%.

Human homeothermy is maintained by both physiological and behavioural control (see *Electrotherapy Explained*, Chapter 7). Human beings seem to have evolved a particularly efficient heat loss mechanism. The combination of a largely hairless skin and an extensive and vigorous sweating response means that quite high external temperatures can be tolerated for prolonged periods provided water is freely available. It can be seen that the skin and subcutaneous tissue have the major role in thermal control.

A further important consideration is the ratio of volume to surface area. The amount of heat energy generated is a function of the volume of the organism, whereas heat loss is largely dependent on the surface area. Large animals like the elephant and hippopotamus have difficulty achieving sufficient rates of heat energy loss and have developed special mechanisms or behaviours to cope with this problem. Thus elephants have greatly enlarged ears to increase their surface area for heat loss and hippos constantly immerse themselves in water or mud. Small homeothermic animals have the opposite problem in cold environments, having to maintain an exceptionally high metabolic rate to keep their temperature up and thus needing to eat relatively large quantities of food. These factors also apply to humans. Babies are theoretically more at risk from heat energy loss than adults but, perhaps, their ability to mobilize brown adipose tissue to add thermal energy compensates for this. This fatty tissue, brown because of its vascularity, is able to metabolize and thus release heat energy directly in response to adrenaline in infants. Its value in adults is not certain. Certainly the survival of the newborn after accidental cold stress has often been noted. Peoples living in tropical countries are considered to be able to survive better if they are tall and slim, i.e. have a relatively high surface area to volume ratio.

As has been noted already, because the body is largely water which has a high specific heat, it takes a good deal of energy to raise the body temperature (see page 153). Thus there is considerable thermal inertia. Furthermore, small temperature changes are acceptable for short periods so that the body effectively stores heat energy during vigorous exercise for dissipation later. Large animals in hot environments, such as camels, seem to be able to allow their body temperature to rise several degrees during the very hot day, losing heat energy at night to restore a normal temperature.

Both heating and cooling the body can be effected by using outside sources of energy. Thus standing close to a fire or radiator adds to body heat energy by the increased amount of absorbed infrared radiation. Local tissue heating is, of course, a widely applied physiotherapeutic measure and includes conduction heating (see Chapter 8 of *Electrotherapy Explained*) as well as radiations (see Chapters 12 and 13 of the same). Therapeutic local heating is also achieved by introducing some form of energy which is transmitted through the tissues to be absorbed and generate heat during its passage, i.e. diathermy. These include shortwave diathermy, microwave diathermy and ultrasound (see *Electrotherapy Explained*, Chapters 10, 12 and 6 respectively). These methods of heating are sometimes called conversion heating because the energy is converted to heat energy in the tissues. The significance of the diathermies is that they are able to deliver heat energy to the deeper tissues because they can bypass the insulating effect of the subcutaneous tissues to some extent. Radiations and conduction heating substantially only cause surface effects.

Local cooling also has therapeutic value (see *Electrotherapy Explained*, Chapter 9) and is usually effected by applying melting ice to the skin surface. This lowers the skin temperature as heat energy is used in melting the ice (see page 157). Evaporative sprays are used for the same purpose. The rapid vaporization of some liquids, such as ethyl chloride or fluorimethane, uses energy from the surroundings, thus cooling any surface to which it is applied.

8.   *Electromagnetic radiations*

Fundamental forces and fields
  The electric field between charges
  The magnetic field
  Electromagnetic radiation
Nature of electromagnetic radiation
Wavefronts and rays
Velocity of electromagnetic radiations
Interactions of electromagnetic radiations with matter
Electromagnetic waves at boundaries
  Reflection
  Relationship between reflection and penetration
  Refraction
  Diffraction
  Penetration and absorption
The emission of radiation from matter
  Energy levels and lined spectra
The production of electromagnetic radiations
  Radio waves
  Microwaves
  Infrared radiation
  Visible radiations
  Ultraviolet radiations
  X-rays
  Gamma rays
  Cosmic rays
  Fluorescence
  Lasers
  Measurement of laser output

FUNDAMENTAL FORCES AND FIELDS

Earlier in this book consideration has been given to the structure and
nature of matter as well as some of the forces involved. It seems
appropriate to try to draw some of this together and set electromagnetic
radiations into context.

   There are only four fundamental interactions in the universe. By far
the most readily recognized is *gravity*. Every particle that has mass has a
gravitational field that extends outwards in all directions and decreases
in strength in proportion to the square of the distance from the centre of
the particle. Since this is an incredibly small force it is insignificant when
considering single particles like atoms but becomes evident when
enormous numbers of particles are collected together to form an object
of large mass. Thus the earth exhibits a well-recognized gravitational
pull on all objects on its surface which is proportional to their mass,
familiar as weight. The moon exerts a lower force proportional to its

smaller mass. Gravitational force accounts for the observed pattern of movement of the suns and planets of the universe.

Particles that have an electric charge exhibit an *electromagnetic* field which also decreases in proportion to the square of the distance from the particle. The electromagnetic field of a given particle is immeasurably stronger than its gravitational force but, as noted, is of two opposite kinds of electric charge, positive and negative. Thus in most situations the presence of equal numbers of opposite charges leads to equalization and elimination of the electromagnetic force, in atoms, for instance, where the positive protons are equalled by the negative electrons. An imbalance between the two kinds of charge is the reason for an electrical force or voltage and the cause of electric charge movement, as has been discussed in Chapter 4. The very strength of the electromagnetic field makes its presence evident on a much smaller scale than that of gravitational fields. Electromagnetism is responsible for the structure, behaviour and chemical properties of atoms and for electromagnetic radiations.

The two remaining fields are entirely confined to the nucleus of the atom. Thus hadrons (mesons and baryons) are the source of a field that spreads out but diminishes in strength so rapidly that its only effective area of influence is within the atomic nucleus. It is over a hundred times the strength of the equivalent electromagnetic field but, of course, only in the atomic nucleus. It is therefore called the *strong interaction* or *strong force* and is largely responsible for holding the protons and neutrons together in the atomic nucleus. The other intranuclear field is due to a particle called the lepton and is also confined to the nucleus. It is very much weaker than the equivalent electromagnetic field (about one hundred-billionth of it, in fact). It is therefore called the *weak interaction* or *weak force*. Although much weaker than either the electromagnetic or the strong force it is still enormously stronger (about 10^{28} times stronger) than the gravitational force in the nucleus. It is responsible for the behaviour of subatomic particles.

The word field has been used in the above description to indicate the region in which the gravitational or electromagnetic forces can exert an influence. Electric or magnetic fields were noted in Chapter 4. They are all really ways of representing – modelling – the way in which a force exists between two objects or bodies. The four interactions just described account for all particle and, through them, all physical behaviour ultimately.

Electromagnetic phenomena have already been considered in Chapter 4 and it can be seen that there are three interrelating aspects:

1. The electric field between charges.
2. The magnetic field.
3. Electromagnetic radiation.

The electric field between charges

There is an attractive force between any negative and any positive charge. The electron is the negatively charged particle and the proton the positive. Similarly, there is a repulsive force between like charges. This is called *electrostatics*. The electrostatic field can be visualized by drawing lines connecting the positive and negative charges, as has been done in Figure 4.9. This is quantified in what is known as Coulomb's law which states that the force between two charges is directly proportional to the magnitude of the charges and inversely proportional to the square of the distance between them. (This was first described in 1785 by Charles Augustine de Coulomb.) This is expressed in the equation:

$$F \propto \frac{q_1 q_2}{d^2}$$

in which F is the force, q_1 and q_2 are the two charges and d is the distance between them. This is exactly the same form as Newton's law of gravitation. The force of gravity, F, between two bodies of mass, m_1 and m_2, separated by a distance, d, is:

$$F \propto \frac{m_1 m_2}{d^2}$$

In practice, electric fields are often described by the effect they have on a test charge placed at a point in the field. This can define the force per unit positive charge and thus describe the intensity and direction of the field at that point.

Electric lines of force indicate the direction of the field – the direction in which a charge would move if free to do so. By convention, the strength of the electric field is denoted by the number of lines passing through unit area. The closer the lines are together, the stronger the force in this region of the field. Thus a single positive charge would be represented by a point with lines radiating away from it equally in all directions, indicating a field uniformly diminishing in strength with increasing distance. Since the field is enlarging with distance in three dimensions, not just the two illustrated on a flat paper surface, it will diminish in intensity proportional to the square of the distance, as already noted. Similarly, a single negative charge would show a field illustrated in exactly the same way except the lines would be shown travelling towards the point. A more realistic situation is described by the electric field existing between positive and negative charges. This is similar to the idea illustrated in Figure 4.19, even though this is showing a magnetic field. When the force (measured in newtons) on the test charge is measured, it gives the field intensity per charge (in coulombs) at that point. So the field intensity is in units of newtons per coulomb or in volts per metre if the rate of change of electrical potential with distance is considered. The numerical value is the same in both cases.

The concept of a field, in the sense described above, is widely used and understood. For example, the familiar weather map, in which barometric pressure is plotted as a series of lines joining points of equal pressure, isobars, is showing a field. Where the lines are close together there is a steep gradient of pressure; when far apart the pressure is more nearly uniform over a large area. Note that this is a scalar field. A vector field, like the electric field, is one that shows direction as well as intensity. In the weather map context, a map showing wind direction and strength would be illustrating a vector field. It is important to understand that electric lines of force are a means of illustrating a force in space, the lines only representing the forces.

The magnetic field

The magnetic field arises only when electric charges move. Magnetism is an aspect of electric force which occurs only when charges move. An existing magnetic field can only affect an electric charge if they move relative to one another. Now, when electric charges move they do so in the line of their electric field, i.e. along an electric line of force as considered above. The resulting magnetic field is at right angles to the direction of motion. This has been considered in Chapter 4 and Figures 4.19 and 4.20 illustrate the magnetic field of a bar magnet and an electric current respectively. The magnetic field is a consequence of charge motion and this motion must have direction in one dimension, such as electrons moving in a wire. As the magnetic force is at right angles to this direction then it will be spreading in two dimensions. Hence the strength of the magnetic field will be inversely proportional to the distance from its source, not to the square of the distance, as in the electric field.

A further feature of the magnetic field is the fact that it has polarity, i.e. it acts in a particular direction. Due to the way in which an understanding of magnetism developed historically, the directionality is described in terms of north and south poles (see Chapter 4). If two loops of wire carrying current in the same direction are placed close to one another the two magnetic fields produced are attracted to one another and will contribute to a combined larger field. If the currents are in opposite directions the magnetic forces will repel one another, producing a combined magnetic field at the centre of which there will be no effective magnetic force. When the loops of wire are placed at right angles to one another, instead of parallel, the magnetic forces will also be at right angles and have no effect upon one another; see the discussion on electric motors in Chapter 4.

Magnetic fields due to a bar magnet are often investigated with the use of iron filings, producing a pattern like that of Figure 4.19. The steel bar behaves as a magnet because electron movement in ferromagnetic materials is aligned in such a way as to summate the individual magnetic forces instead of each being cancelled out, as occurs in the random electron motion of most materials. This is explained on page 65. The filings act like individual test charges, described in connection with the electric field. The same effect can be produced by marking the position of a small free-swinging compass needle in different parts of a magnetic field. The iron filings act as tiny individual magnets; the pattern into which they fall is determined by the direction and strength of the magnetic field. The resultant linear pattern illustrates the magnetic field expressed in magnetic lines of force. Again it is a vector field and again the lines are representative of the real forces. Whereas the electric field is based on point charges, isolated magnetic poles do not exist so that the magnetic field is described by its effect on a dipole, the tiny iron filing or compass needle, which has both north and south poles.

The strength of a magnetic field can be described either as the magnetic flux density (often denoted B) or the magnetic field strength (often denoted by H). Both of these are vector quantities and are distinct because the flux density includes the magnetic permeability of the medium in which the field is acting. Thus:

$$B = H \times \text{magnetic permeability}$$

The word flux has special meaning in this context. Originally magnetic field theory was developed with analogies connected with fluids so that flux in the meaning of a flow or flowing came to be used. This is somewhat confusing since there is nothing that flows in a magnetic field. Magnetic flux is measured in webers (after Wilhelm Weber, 1804–1891, who published a unified theory of electric and magnetic forces in 1846). The tesla is the unit of magnetic flux density and is 1 weber per square metre. It is a measure of force because it is the equivalent of newtons per ampere per metre. (Nikola Tesla, 1856–1943, pioneered the development of electrical machinery in North America.)

A magnetic field, then, occurs as a result of moving electric charges and is proportional to them. Note that a steady electric current produces a steady magnetic field and a varying current produces a varying magnetic field. Similarly, a varying magnetic field can induce a varying current (see electromagnetic induction, Chapter 4) but a steady magnetic field has no effect.

Electromagnetic radiation

Faraday was instrumental in developing the concepts of electric and magnetic fields as a means of visualizing electric and magnetic phenomena. James Clerk Maxwell (1831–1879), the Scottish physicist, took Coulomb's law, the fact that currents produce magnetic fields, that isolated magnetic poles do not exist, and magnetic induction to formulate mathematically four equations which effectively synthesized electromagnetic interactions. From these he was able to predict that changing electric and magnetic fields should produce a wave of electromagnetic energy propagated through space at a velocity of 3×10^8 m s^{-1}.

To see how this might happen, consider a rod carrying an oscillating current so that one end is made negative when the other is positive and constantly reverses as electrons are accelerated to and fro. This produces an oscillating magnetic field around the rod. Now changing magnetic fields give rise to induced currents – electric fields – in conductors and even in space. (This is what Maxwell was able to show.) These changing electric fields give rise to changing magnetic fields which, in turn, produce changing electric fields and so on. The two fields are associated in this way so that waves of varying electric and magnetic fields are propagated through space. The magnetic field would spread outward around and away from the rod, setting up an electric field at right angles to it. Thus, these electromagnetic radiations radiate out into space with the electric and magnetic fields at right angles to one another and both at right angles to the direction of travel (Fig. 8.1). As the variations in the fields are sinusoidal and at right angles to the line of propagation it is therefore described as a transverse wave (see Chapter 3). From what has been noted it will be understood that these electromagnetic radiations are waves in the sense that variations in electric and magnetic fields occur in a wave-like manner but no material – matter – is involved. Thus these radiations will travel indefinitely through space. On encountering matter, radiations may be able to pass through or may be absorbed, having their energy changed to some other form. They may also have their direction changed and be reflected from the surface of matter. What happens depends on the type of radiation and the nature of matter encountered.

Once generated, electromagnetic radiations are independent of the moving charges that provoked them in the first place. Like all waveforms, these radiations convey energy but they do so in a special way, in that the energy is in discrete units called photons or quanta (discussed later). Except when there is some interaction with matter (see page 193), these radiations travel in straight lines. Thus the radiation emitted from the rod, considered above, would spread out in all directions. All these radiations travel at the same velocity but differ in their wavelength and frequency. It is this variation that determines the kind of radiation. Electromagnetic radiations are most familiar as light but are in exactly the same form as radio waves or X-rays except for their wavelength and frequency. The full list of electromagnetic radiations is shown in Table 8.1.

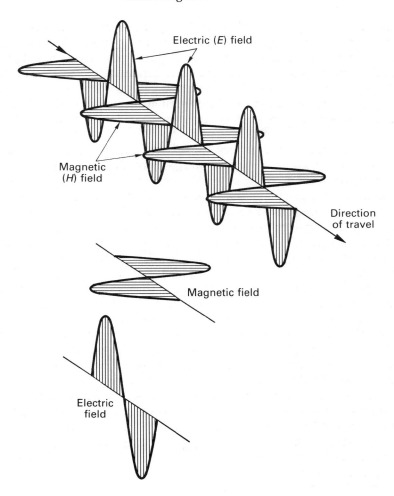

Fig. 8.1   Electric and magnetic fields.

To summarize what has been observed so far: the four fundamental interactions in the universe are, in order of their strength within the atom:

1. The *strong force*, which acts only in the nucleus.
2. *Electromagnetism*:
   (a) Electrostatic fields – the attraction and repulsion of stationary charges.
   (b) Magnetic fields – the consequence of moving charges.
   (c) Electromagnetic radiation – the consequence of acceleration and deceleration of charges.
3. The *weak force*, which acts only within the atomic nucleus.
4. *Gravity*, which acts at distances on massive objects.

Table 8.1 The electromagnetic spectrum

Radiation	Subdivisions	Wavelengths	Photon energies
Radio waves	Long-wave Medium-wave Short-wave and television	3–1 km 600–180 m 30–3 m	4.1×10^{-11} eV to 1.2×10^{-2} eV
Microwaves		1 m–1 cm	
Infrared	Long Short	100 μm–2000 nm 2000–770 nm	1.2×10^{-2} eV to 1.8 eV
Visible	Red Orange Yellow Green Blue Indigo Violet	750–650 nm 650–600 nm 600–550 nm 550–500 nm 500–470 nm 470–440 nm 440–400 nm	1.8–3.1 eV
Ultraviolet	A B C Short 'vacuum'	400–315 nm 315–280 nm 280–100 nm 100–10 nm	3.1–124 eV
X-rays	Soft Hard	10–0.1 nm 0.1–0.01 nm	124 eV to 124 MeV
Gamma		0.01 pm	

In many cases the wavelengths given are examples and do not define the limits of the particular subdivision.
Photon energies from Hay and Hughes (1972).
N.B. nm or nanometre = 10^{-9} m; pm or picometre = 10^{-12} m (see Appendix E).

NATURE OF ELECTROMAGNETIC RADIATION

From what has been described so far it is clear that electromagnetic radiations can be described in terms of wave motion (see Chapter 3). Other types of waves have been considered, notably sonic waves, which are longitudinal waves – discussed in Chapter 6. They differ from electromagnetic waves because they involve displacement of the atoms and molecules of the medium in which they pass, e.g. compression and rarefaction. This controls the velocity at which these waves can progress. Electromagnetic waves are variations of electric and magnetic fields in space, as explained, and are therefore independent of atoms and molecules. They travel at a fixed velocity of approximately 3×10^8 (2 997 925) m s^{-1}. This important figure is variously expressed as 300 million m s^{-1} or 300 000 km s^{-1}. This is the velocity in empty space or *in vacuo* and is expressed by the symbol *c*. For most of the types of radiations the velocity in air is little different.

In Chapter 3, the relationship between the velocity, wavelength and frequency of waves was discussed and the general relationship

$$\text{Velocity } (v) = \text{wavelength } (\lambda) \times \text{frequency } (f)$$

was expressed. For electromagnetic waves, since the velocity is constant and known, information of either wavelength or frequency allows the other to be deduced. The wavelengths extend from hundreds of kilometres to fractions of a picometre and frequencies of a few hertz to millions of terahertz (Tables 8.1 and 8.2). Appendix E may be helpful in clarifying the units of length.

It can be seen from Table 8.1 that the distinction between some very different forms of radiation is only a small difference of wavelength and hence frequency. One striking feature is the very small size of the collection of radiations recognized as visible light, from about 390 to 760 nm. Yet the eye is able to distinguish differences in wavelength of a few hundred nanometres to recognize the different colours quite distinctly. The collection of electromagnetic radiations resolved into their frequencies and wavelengths in this way is called a spectrum. The visible radiations or spectra have been widely studied simply because they are visible. When certain substances emit groups of particular visible radiations, they are called emission spectra.

Table 8.2 Some examples of electromagnetic radiations to show the relationship of frequency to wavelength and where they are found

Frequency	Wavelength	Where found
1 kHz	3×10^5 m or 33 km	
1 MHz or 10^6 Hz	3×10^2 or 300 m	In medium-wave radio band
27.12 MHz	11.06 m	Shortwave therapy
1 GHz or 10^9 Hz	300 cm	Microwave region
2.45 GHz	12.245 cm	Microwave therapy
1 THz or 10^{12} Hz	300 μm	Infrared region
30 THz or 3×10^{13} Hz	10 μm	Peak infrared emitted by body
300 THz or 3×10^{14} Hz	1000 nm	Approximately maximum penetration depth for short infrared in tissues
600 THz or 6×10^{14} Hz	500 nm	Green visible radiation
1000 THz or 10^{15} Hz	3300 nm	Ultraviolet B
10^{17} Hz	3 nm	X-ray region

Table 8.1 may be superficially misleading because it shows radiations by description and gives no graphic indication of the extent of the total spectrum occupied by particular groups of radiations. Table 8.2 may help in this respect. The visible radiations occupy only a tiny fraction of the total electromagnetic spectrum. One confusing feature of electromagnetic radiations is the fact that radiations were described and utilized in different contexts so that they became described by custom in different ways. Thus radio waves (also called Hertzian waves) were originally divided into bands by their wavelength, as indicated in Table 8.1. Subsequently they were designated by frequency, the grouping being by international agreement into these radio frequency bands (Table 8.3). Microwaves are also known as radar, which is an acronym of *r*adio *d*etection *a*nd *r*anging. This came about because the principal use of microwave radiation and the reason for its development is the detection of aircraft, ships and missiles.

The familiarity of and ability to recognize and measure visible radiations have meant that much of the behaviour of other electromagnetic radiations is understood and illustrated by reference to experiments done with light. The naming of infrared, below the red, and ultraviolet, beyond the violet, indicates their relationship to the colours of the visible spectrum and how visible light was taken as a starting point. The point is further illustrated by the fact that the limits of the visible spectrum – determined by what the human retina can detect – are about the only two agreed points that are universally fixed. There are arbitrary divisions between the other types of radiation, being sometimes supported by international agreements. The naming covers wide bands of radiations which do not necessarily exhibit the same effects, hence some of the further subdivisions shown in Table 8.1 – ultraviolet divided into A, B and C or X-rays into hard and soft, for example. There is also some overlap between many of these bands of radiation, such as between microwaves and infrared radiations or X-rays and gamma rays. In some instances the naming of the radiation depends on how it has been produced rather than on its frequency and wavelength. It will have been noticed that electromagnetic energy is being described both as electromagnetic waves and as electromagnetic radiations. This is deliberate as

Table 8.3 Radio frequency bands

Description	Frequency
Very low frequency	Below 30 kHz
Low frequency	30–300 kHz
Medium frequency	300–3000 kHz
High frequency	3–30 MHz
Very high frequency	30–300 MHz
Ultra high frequency	300–3000 MHz
Super high frequency	3000–30 000 MHz
Extremely high frequency	30 000–300 000 MHz

both terms are appropriate and widely used, as well as to emphasize the concept considered below.

From a consideration of what has been discussed about the strength of electric and magnetic fields and the behaviour of waves, described in Chapter 3, it would be reasonable to expect that the energy carried by electromagnetic radiation would depend on the energy of the source. This is true up to a point and when the photoelectric effect was discovered by the German physicist, P. Lenard, it was expected that the number and energy of electrons emitted from a metal plate would correspond to the intensity of illumination. This turned out not to be the case, since some radiation – red light for example – applied to some metals had no effect whatsoever, no matter how intense the radiation. On the other hand quite low intensities of blue light would 'kick' electrons off certain metal plates. Other incongruities, concerning the radiation emitted from heated bodies, had led the German physicist Planck to suggest, in 1900, that electromagnetic radiation consists of small separate units or packets of energy which he called quanta. (*Quantum*, the singular, is the Latin for how much). He further hypothesized that the amount of energy in each quantum depended on the wavelength of the radiation; the shorter the wavelength, the more energy in the quantum. The quantum energy, e, can be shown to be equal to the frequency, f, multiplied by a constant, h, called Planck's constant. Thus:

$$e = hf$$

Now h is a very small number, having the value of 6.626196×10^{-34} joule seconds, so that recognizable light sources emit enormous numbers of quanta per second. It has been calculated that a 10 W source emits approximately 3×10^{19} quanta per second. It was Albert Einstein who saw that this explained the paradox of the photoelectric effect. To absorb enough energy to force an electron off the metal surface it must be struck by a unit of radiation of sufficient size – a quantum. If the electron is weakly held to its parent atom then low-energy quanta will work but if it is held more firmly, then higher-energy quanta, i.e. those of shorter wavelength, are needed. (For his work in connection with the photoelectric effect, Einstein received the Nobel prize in 1921. His Special Theory of Relativity, presented in 1905, is an extension of quantum theory, suggesting that all electromagnetic radiation travels through space in quanta.) A little later the same 'photon' was given to this quantum unit and thus in some sense electromagnetic radiation can be regarded as a stream of particles. The photon can also be regarded as an elementary particle (see Table 2.3).

There would now seem to be two sets of properties pertaining to electromagnetic radiation – those of a waveform and those of a particle. This is sometimes seen as a paradox, the wave–particle duality, but it is really two aspects of the same phenomenon. The difficulty of accepting that electromagnetic radiations are at one and the same time a waveform and a particle is largely due to human preconceptions of the nature of waves and particles. Waves on the surface of water are easily seen and the compression of air molecules in sound waves, for example, can be readily visualized. In these examples, matter is being seen or imagined but electric and magnetic fields are regions of varying force – no physical entity is being thrown into waves. The waves can be thought of as a way of describing the probable path of a photon. With many millions and millions of photons in even the weakest radiation the average position of photons can be described as waves but the path of a single photon cannot be detected or predicted. Due to describing photons as particles, the image of something like a solid ball is generated, when an equally valid description of photons might be a unit of energy. When electromagnetic radiations are travelling, it is the wave motion that is considered and which describes their behaviour. When they are being emitted or absorbed, it is the particle – photon – activity that best defines their behaviour. Photons are the result of interactions with single electrons and when absorbed cause energy transfers to single electrons; thus they are described as discrete packets of energy. In the photoelectric effect described above, photons with insufficient energy were unable to release an electron from one metal, no matter how many bombarded it. But photons with higher energy, provided by radiations of higher frequency, each have sufficient energy to dislodge an electron, even in small numbers. A similar homely example is provided by photographic film which is immediately darkened by exposure to visible or short-wavelength infrared radiation and yet totally unaffected by longer-wavelength infrared. This is evidenced by the fact that the film must be handled in darkness but the infrared from the warm hand holding it does not fog the film.

The photon energy can also be expressed in relation to electrons as the electron-volt (eV). This unit is widely used in nuclear physics and in connection with radioactivity. The electron-volt is the increase in energy of an electron when it is accelerated through a potential difference of 1 V. One electron-volt is equal to 1.602×10^{-19} J. Table 8.1 gives some electron-volt energies for each group of radiations.

Electromagnetic radiations also differ in other ways. Most radio waves are polarized, whereas the rest are non-polarized. This occurs because radio and microwaves are usually generated by oscillating electrons in a wire in one direction or plane, usually vertical or horizontal. The resulting electric field is in the same plane, with the magnetic field at right angles and the radiowaves are said to be plane-polarized. To be an efficient receiver the aerial or antenna needs to be in the same plane; television aerials are either vertical or horizontal. Infrared, visible and other radiations are usually generated by processes of random electron movement, which means that the electric fields of the radiations can be at any angle perpendicular to the line of travel of the radiation – some vertical, some horizontal and some at every angle between these (Fig. 8.2). Thus the electric fields, E fields, are oscillating in all possible

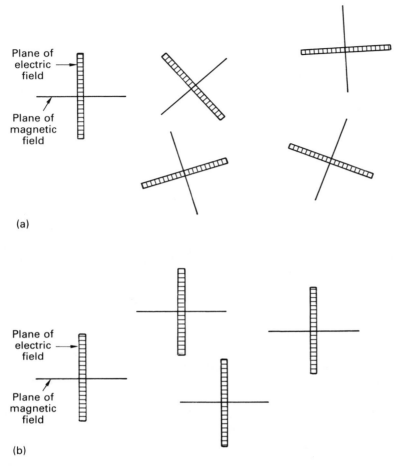

Fig. 8.2 Polarization of radiations: (a) non-polarized radiation – fields randomly oriented; (b) plane-polarized radiation – the electric field is vertical.

planes. If visible radiation is passed through certain materials such as polaroid (which is a polyvinylalcohol impregnated with iodine and stretched during manufacture) only radiations in a certain plane are allowed to pass through. The light is said to be polarized. This occurs because the molecules are aligned in parallel to each other due to the stretching and absorb radiations whose electric field is in the same plane but do not affect electric fields at right angles. This is the principle on which polaroid sunglasses work. Much of the glare is in the form of reflected radiations from horizontal surfaces – a wet road, for example – so the polaroid glasses are oriented to absorb the horizontal electric field (Fig. 8.2). In many circumstances light is partially plane-polarized rather than totally polarized. Passing polarized light through certain materials may allow the arrangement of their molecules to be investigated. Muscle tissue is an example in which the long chains of protein molecules absorb polarized light in one direction more than the other, thus displaying their regular structure.

Since the magnetic and electric fields of a radiation are at right angles to one another, illustrated in Figure 8.1, it may be asked which field is affected by polarizing materials. The answer is the electric field, the E field. It is the E field which acts on the retina and is absorbed by the metal of aerials, the magnetic field being unaffected.

The millions upon millions of photons are being emitted at random times so that the peaks and troughs of their waveforms do not coincide. Further it has been noted already that radiations from most sources tend to be emitted in all directions. This means they are oriented at random to one another about their axis in the line of propagation (Fig. 8.2).

It has also been indicated that different photons will have different energies depending on their frequency. There are circumstances in which radiations can be generated so that they consist of identical photons in step with one another and all travelling in the same direction. This will be considered later as laser radiation (see page 212).

WAVEFRONTS AND RAYS

The manner in which electromagnetic radiations or waves travel in space and how they interact with matter can be looked at in two different ways depending on whether the features are best described as a stream of particles or a waveform. Thus they can be illustrated by drawing rays or wavefronts.

A ray is a line drawn to represent the path taken by a radiation from source to destination. Bundles of rays can be drawn so that their number per unit area is representative of the intensity of the radiation, i.e. rays can be drawn spreading out from a point source or converting through a lens. Thus rays can illustrate more than simple direction (Fig. 8.3).

Wavefronts are lines drawn joining the same points on parallel waves, thus these lines might represent the crests or troughs of the waves. In fact waves in water, say waves rolling up a beach, are simply two-dimensional wavefronts made visible. Figure 8.3 shows both rays and wavefronts and indicates that they are at right angles to one another. It must be understood that these are drawn in two dimensions but are intended to represent radiations emerging in all directions. While the scattering of dust particles in the path of a beam of light gives a visual impression of a ray of light, the behaviour of waves can only be visualized on, say, a water surface.

Electromagnetic radiations will travel in space in straight lines. This is obvious in the case of visible radiations since shadows are formed with sharp edges where the path of the light is intercepted by an object. For other radiations in the electromagnetic spectrum it is true in space, but longer radio wavelengths are able to pass around solid objects, like sound waves (see Chapter 6). This property of rectilinear propagation

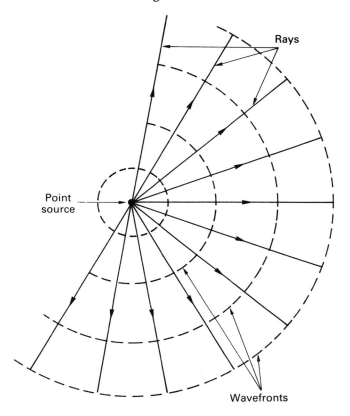

Rays

Point source

Wavefronts

Fig. 8.3 To illustrate electromagnetic radiations travelling away from a point source by rays and wavefronts which are perpendicular to one another.

and the small source shown in Figure 8.3 allows this figure to show how the intensity of the radiation – the energy through unit area in unit time – will diminish with distance. Since the radiations spread out in all directions the intensity will fall in proportion to the square of the distance. This is expressed in the *inverse square law* which states that the intensity of radiation from a point source is inversely proportional to the square of the distance from the source. Thus:

$$I \propto \frac{1}{d^2}$$

where I is the intensity and d the distance.

The reason for this relationship is illustrated in Figure 8.4. Although in real situations point sources of radiation are rare and other factors may modify the strict application of this principle, it is none the less of great practical importance in the application of infrared and ultraviolet radiations for therapy. Inspection of Figure 8.4 shows how small changes of distance will lead to large changes in intensity.

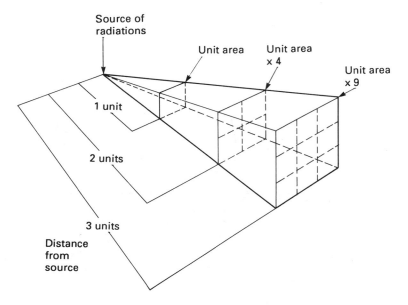

Fig. 8.4   The inverse square law. The intensity of radiation from a point source is inversely proportional to the square of the distance from the source.

VELOCITY OF ELECTROMAGNETIC RADIATIONS

As the speed of radiations *in vacuo* has already been emphasized as a constant of nearly 3×10^8 m^{-s}, it may be wondered what happens in the more usual situation of the passage of radiations through air or other transparent material. This can be established by measuring the refractive index (see page 200), which can easily be done for visible radiations. The index of refraction of a substance is the ratio of the velocity of light in a vacuum to the velocity in that substance. Table 8.4 gives some examples. The speed of radiations *in vacuo*, *c*, is the highest attainable and so speeds in any other substances are bound to be lower. The velocities of radiations of different wavelengths differ in the same material. The reason why radiations are slowed in matter is that they interact with the electric fields of the molecules and atoms of the material concerned. It will be recalled that at ordinary temperatures the atoms and molecules are in constant motion. Some molecules will have a negative electric charge at one end and a positive charge at the other and are said to be polar. If an electric field is applied first in one direction and then in the other they will rotate to and fro. The electron clouds of non-polar molecules and atoms will be disturbed or distorted by the electro-magnetic field of the waves so that they are, on average, nearer one end than the other. Such molecules are said to be polarized, the polarization following the alternating field. Any charged particles, electrons or ions, will also be affected and will move or try to move in the direction of the

Table 8.4 Velocity of yellow light in various substances

Substance	Index of refraction	Velocity of light
Air	1.0002926	2.9912×10^8 ms^{-1}
Glass (particular type)	1.5	2×10^8 ms^{-1}
Water	1.333	2.25×10^8 ms^{-1}
Diamond	2.4168	1.2413 ms^{-1}
Crystalline quartz	1.553	1.9317 ms^{-1}
Ethanol	1.36	2.2059 ms^{-1}

field. Any movement of particles can transfer energy from the wave to the material. Regular motion in the wave is converted to random motion in the structure of the material. For example, electromagnetic energy is converted to heat energy in the material through which it travels.

INTERACTIONS OF ELECTROMAGNETIC RADIATIONS WITH MATTER

Radiations from the sun enter the earth's atmosphere and some radio-waves, infrared, visible and long ultraviolet pass almost unhindered through the gases of the air. All radiations, when they meet matter, exhibit three possible interactions:

1. The radiations may be turned back at the surface; they are *reflected*.
2. The radiations may pass through the material, in which case they are *transmitted* and the material is said to be transparent to the radiations. Transparency refers to light waves being able to pass through a substance, as they can through glass or clear water and by extension to other radiations, for instance, wood and bricks are largely transparent to radiowaves. Once radiations enter a material they are said to have *penetrated* it but in doing so their line of travel can be altered. They may be bent or *refracted* in one direction or parts of the radiation may be reflected or refracted in the material so that the radiation is said to be *scattered*.
3. Having entered the material the radiation may be *absorbed*. That is, the electromagnetic energy of the radiation is converted into some other form of energy. The vast majority of such interactions involve heat energy being produced in the material, as already noted. Other familiar conversions are the production of electrical signals, as occurs in radio or television transmission, and the production of carbohydrates in plants by photosynthesis. There are innumerable other energy conversions but for any to occur the radiation must be

absorbed. In other words, if the radiation is reflected from or transmitted through the material then it has no effect on the material. This idea was first suggested by Grotthüs in 1820; he said that only the rays absorbed are effective in producing chemical changes; hence this is referred to in older texts as *Grotthüs's law*.

In almost all real situations all three – reflection, absorption and penetration/transmission – occur together but in different degrees. What happens depends on the nature of the material and the frequency/ wavelength of the radiations. For example, light will pass readily through panes of window glass but some of the light will be reflected. If this were not so the glass would be invisible. Also some small amount of the radiations is absorbed. On the other hand a black matt surface (such as a black mat!) is an almost perfect absorber of visible radiations whilst a shiny metal surface can be a near perfect reflector. While these are obvious because they involve visible radiations, the same thing occurs with other types of electromagnetic radiations. Thus some microwave frequencies are strongly reflected from metal surfaces but will penetrate and be absorbed in human and animal tissues. This is what happens in a microwave oven in which the radiations are reflected from the metal walls and absorbed, mainly by the water, in food, hence heating it. Infrared radiations are strongly absorbed by the skin surface, as noted earlier, but reflected from polished metal surfaces. Ultraviolet B is transmitted quite well through water, quartz and certain special glass but largely absorbed or reflected from ordinary window glass. Very short ultraviolet radiations are absorbed by nearly all matter, hence called vacuum ultraviolet. X-rays are well known for being able to penetrate less dense tissue but not bone, hence providing an X-ray picture of the skeletal structures. The most obvious and visible evidence of selective reflection and absorption by different surfaces for different wavelengths is provided by the recognition of colour. The colour blue is due to reflection of more wavelengths near the blue end of the colour spectrum and absorption mainly of those wavelengths at the red end. It is not usual for all the wavelengths other than blue to be absorbed entirely; rather, there is a predominance of reflected blue light.

Colours are, of course, perceived in other ways as well as their wavelength. The three attributes recognized are hue, which is colour recognized by the mix of wavelengths; brightness, which is the intensity; and saturation, which is, roughly speaking, the purity or density of the colour – another consequence of the mix of wavelengths. Colour is also perceived in transmitted light which is seen, for example, in looking at a stained glass window inside a church or cathedral. In this case the colours are seen because more of the blue wavelengths around 400 nm are transmitted and other wavelengths are absorbed. The colours just discussed differ from one another by only a few tens of nanometres. The cells of the retina are extremely sensitive to very low energy levels and can detect radiation of only a few photons under certain circumstances.

The light-sensitive cells of the retina, the cones, each contain one of three pigments which differ in that they absorb different wavelengths. The particular groups of cones excited by different wavelengths have specific connections in the brain which thus allow colour to be recognized. This is, of course, an example of electromagnetic energy being converted to chemical energy in the retinal cells. If all the wavelengths of the visible spectrum are present at equal intensities then the resulting perceived light is said to be white. Thus the radiations reflected from this page are of all visible wavelengths in equal proportions except where the print – the black ink – is absorbing all visible radiations and reflecting none.

Other mammals and birds seem to have vision similar to humans but insects have a rather different mechanism. The insect eye is made of a series of separate transparent fibres, each transmitting light which strikes it straight on only. These fibres, or ommatidia as they are called, are arranged in the rounded compound eye of the insect in different directions so that objects must be detected by the number and pattern of ommatidia that they stimulate. A further important difference is that some insects are apparently able to distinguish the pattern of polarization of the scattered, visible radiations of the sky. Visible light is reflected by molecules in the atmosphere, which causes the radiations to be partially polarized, the direction of polarization depending on the position of the sun. It is this mechanism that allows bees to navigate between their hives and sources of nectar (von Frisch, 1967). This means that the bees can navigate even when the sun is hidden by cloud, providing they can see some patch of sky. Thus insects are able to see and use features of these radiations totally invisible to humans.

ELECTROMAGNETIC WAVES AT BOUNDARIES

It has been noted that electromagnetic radiations travel at different velocities in different media. The energy carried by a wave – any kind of wave, not only electromagnetic waves – is a function of the amplitude and the frequency of the wave (see Chapter 3). An increase in either or both means the wave is conveying more energy. But, as has been seen, the nature of the medium through which the wave is passing will influence the form of the wave. For sonic waves it was the density and elasticity of the medium that determined the acoustic impedance. A similar quality, the electrical impedance, depends on the dielectric constant and conductivity of the material. These determine the behaviour of the molecular and atomic electric fields in the material as they are affected by radiation. When an electromagnetic wave approaches the boundary with another medium it interacts with the electric fields of the new material at the same frequency, but unless the impedances of the two media are the same, all the energy of the wave cannot be transferred, so some of it is turned back as a reflected radiation (see Chapter 3).

Reflection

Reflection, therefore, occurs as a result of the difference in impedance of the media in which the radiation is travelling. In many circumstances almost all the energy of the radiation is turned back, such as visible radiations at a polished metal surface. The consequence of a ray, a narrow beam of radiations, applied to a flat reflecting surface is illustrated in Figure 8.5. The direction of the reflected radiation depends on the angle that the original, or incident, ray makes with the reflecting surface. If the reflecting surface is flat, a plane surface, then the angle of incidence is equal to the angle of reflection. Also the incident ray, reflected ray and the normal are all in the same plane, i.e. they are all in the same plane of the page in Figure 8.5. The normal is a line perpendicular to the point of incidence of the ray. These are the laws of reflection and they fully define the direction of the reflected radiation. The laws of reflection are usually illustrated with reference to visible radiation but are equally valid for all types of radiations; the microwaves reflected from the inside of a microwave oven, noted above, for example. Although microwaves penetrate the tissues there is also considerable reflection from the body surface. One of the best microwave reflectors is a metal mesh. The gaps in the mesh are considerably smaller than the wavelength of microwaves used in therapy (12.25 cm) so that reflection from the metal is almost total.

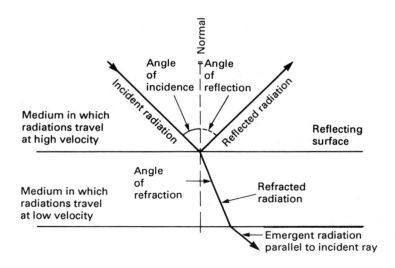

Fig. 8.5   Reflection and refraction.

Being able to control the direction of reflected radiations has many practical uses. Since radiations from a small source, such as a light bulb or heater element, tend to travel outwards in all directions, as in Figure 8.3, placing a curved reflector on one side can cause the radiations to be reflected in a parallel beam in a useful direction. A strictly point source of radiations placed at the principal focus of a small, concave spherical mirror would provide a narrow, parallel beam but a more suitable reflector for a wide beam is a parabolic reflector. In infrared lamps used for therapy, the sources are rather large and extend beyond the focus of the parabolic reflector so that the beam spreads out and is not parallel. This has advantages in that it diminishes the risk of hot spots due to convergence of the radiations and it allows regulation of the intensity by altering the distance of the lamp. This is illustrated in Figure 8.6. The same is true of ultraviolet lamps using small U-shaped tubes. The reflector material is a special aluminium alloy which is a most efficient

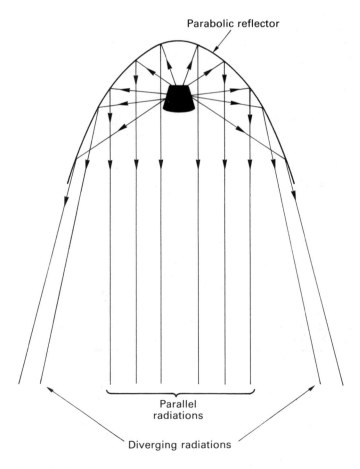

Fig. 8.6 A parabolic reflector.

reflector of utlraviolet radiation. Because it is not such a good reflector of visible radiations it does not look as bright and shiny. Reflectors for radiations being produced from long fluorescent tubes, either for lighting or for ultraviolet therapy, are simply long troughs which act in the same way and can be parabolic in cross-section.

Relationship between reflection and penetration

As noted above, in most circumstances a proportion of radiations are reflected while the rest penetrate and may be absorbed or transmitted. What happens depends on the wavelength of the radiations and the nature of the medium. It also, however, depends on the angle of incidence of the radiations. This is what might be expected from a simple consideration of radiation as a stream of photons. Those that strike the surface at right angles have a greater chance of penetrating than those that strike at any other angle. In fact, the penetration of radiation is proportional to the cosine of the angle of incidence of the radiation. This is often called the *cosine law*. It is illustrated in Figure 8.7. If radiation strikes a surface at right angles the area irradiated is the cross-section of the beam but if the radiation is angled to the surface,

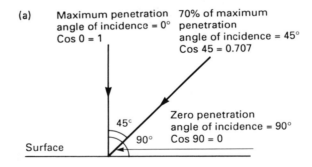

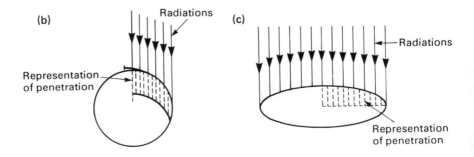

Fig. 8.7 Ratio of reflected and penetrating radiations; (a) the cosine law (b) the parallel radiations applied to a curved surface; (c) parallel radiations applied to an ellipsoid surface.

then a larger area is covered and hence intensity per unit area is reduced. This can be easily seen by shining a torch vertically down to the floor in a darkened room so that a round pool of light is produced. If the torch is now angled, the area illuminated becomes larger and oval and is less bright. Radiations perpendicular to the surface will have an angle of incidence of 0°. The cosine of 0 being 1, this gives the maximum penetration and least reflection, as indicated in Figure 8.7. If the angle of incidence is increased to 45°, the penetration is less in proportion to the cosine of 45°, i.e. 0.7 (Table 8.5), indicating that it is 70% of maximum. At greater angles of incidence the penetration diminishes rapidly, as indicated in Table 8.5, until at an angle of incidence of 90° – when the radiation is travelling parallel to the surface – there will be none, which is an entirely expected result. The cosine of 90° is 0, indicating nil penetration.

There is a second factor which tends to lower the penetration and absorption of a beam of radiations striking a surface at any angle of incidence beyond 0°. The amount of reflection increases with greater angles of incidence but in a rather complex way related to the impedance of the material forming the surface. This may be understood with the analogy of a dart thrown at a dartboard. If it strikes perpendicular to the board, the point is likely to penetrate deeply but if it strikes at an angle there is a greater chance that it will glance off or be 'reflected'.

This decrease of penetration of radiations with greater angles of incidence has important practical consequences. If parallel radiations are applied to a curved surface, such as the surface of the human body, the angle of incidence increases steadily around the curve, as indicated in Figure 8.7. The penetration, absorption and hence the effect of the radiation will diminish around the curve so that any results of the radiation, such as an erythema on the skin due to ultraviolet radiation, will diminish smoothly around the curve. In order to achieve a uniform effect it is necessary to apply radiations from two directions at right angles to one another. In fact, human trunk and limb segments tend to be approximately ovoid in cross-section so that most of the body is more or less evenly irradiated by parallel radiations applied to the front and

Table 8.5 Table of cosines

Angle	Cosine
0°	1.0
15°	0.966
30°	0.866
45°	0.707
60°	0.5
70°	0.342
80°	0.174
90°	0.0

back, e.g. sunbathing in prone and supine lying. It will be realized that ultraviolet radiation is most intense when the sun is directly overhead so that those standing upright tend to suffer sunburn on the upper surface of the shoulders and those wishing an even tan recline to sunbathe. Ultraviolet, infrared and microwave sources used for therapy are usually applied so that the radiations strike the area to be treated at right angles to maximize and standardize the dosage. Inspection of Table 8.5 shows that small variations from the perpendicular are likely to make little difference. In fact, deviations of as much as 25° lead to less than 10% reduction, which is just as well because most of the sources used in therapy have diverging beams of radiation, as shown in Figure 8.6.

Refraction

It has been explained that the velocity of radiations changes in different media so that if the radiation approaches the boundary between the media at any angle other than the perpendicular it is bound to change direction. This is illustrated in Figure 8.5. If we consider the wavefront of a radiation, the part of the wavefront that first enters the new medium travels more slowly and is therefore delayed in comparison to the rest of the wavefront still in the high-speed medium. This causes the wavefront to turn through an angle which depends on its relative velocities in the two media. This is shown in Figure 8.5 in terms of rays showing the changes of direction. As the radiation passes from a medium of low to one of higher velocity it bends again but in the opposite direction, as shown. The terms used in connection with refraction are noted in this figure. For a simple, if sombre, analogy, see Appendix F. While all electromagnetic radiations can be refracted, once again it is usually described in terms of visible radiations.

The laws governing refraction were elucidated in 1621 by Willebrord Snell, the Dutch physicist, who was professor of mathematics at Leiden University. He did not, apparently, publish at the time so that the French philosopher, René Descartes discovered the law independently in 1637.

The first law states that the incident and refracted rays are on opposite sides of the normal at the point of incidence and all three are in the same plane. The second law, known as Snell's law, states that the ratio of the sine of the angle of incidence to the sine of the angle of refraction is a constant for any given pair of media. This constant is called the *refractive index*, denoted n, and it is the ratio of the sine of the angle of incidence (I) to the sine of the angle of refraction (R):

$$n = \frac{\sin I}{\sin R}$$

It is really the ratio of the speed of the radiations in the two media. The velocity of radiations *in vacuo* is taken as 1 and since the velocity is less in all media, the refractive indices are all greater than 1; see Table 8.4 for some examples. As the refractive index is a ratio it has no units. Since the velocity of light in air is very close to that *in vacuo*, the index is often found by passing light from air into the material to be investigated. Of course, the refractive index varies with wavelength and the figures given are for yellow light with a wavelength of 589.3 nm.

The effect of radiation between air and glass has allowed the construction of all kinds of optical instruments. Figure 8.5 shows that radiations entering and emerging from a parallel-sided block of glass are parallel but displaced sideways. If the angle of incidence is steadily increased at the junction between the dense and less dense medium, so that the refracted ray is progressively bent away from the normal, there will come a point at which the refracted radiation will travel parallel to the surface of the medium. This is called the *critical angle*. Any further increase in the angle of incidence will cause the ray to be completely bent back into the denser medium – an effect called *total internal reflection*. This effect is utilized in directing the rays through a pair of binoculars by using prisms. It is also the principle that allows light to be conveyed through narrow glass or plastic fibres, optical fibres, forming light guides, bronchoscopes, arthroscopes and other endoscopes (Fig. 8.8). The fibres used are very fine, of about 10 μm, and are often constructed of two

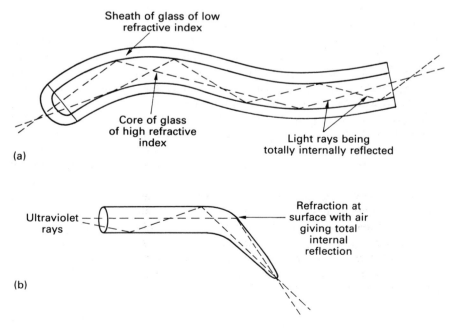

Fig. 8.8   (a) An optic fibre and (b) a quartz rod.

types of glass, the outer glass sheath having a much lower refractive index (Fig. 8.8). While an air–glass junction would be optically better, the need to have a permanent perfect unblemished boundary for refraction is more important. The fibre can be bent to a degree as long as the light strikes the internal surface at an angle of incidence greater than the critical angle. Thus the light travels along the narrow fibre, making hundreds of internal reflections without loss. These devices consist of bundles of fibres and can be arranged so that each individual fibre carries light from a small area of the surface collectively to supply an image at the eyepiece. Thus the arrangement is similar to the ommatidia of the compound eye of the insect. Light can be sent down one group of fibres and reflected from the internal surface back up another.

The eyepiece can, of course, always be exchanged for a camera or television video camera to make a permanent record. Such devices have enabled many medical investigations to be performed non-invasively. Other applications of optical fibres include their use in the transmission of electrical signals for telecommunications, for example. Electrical signals are converted to pulses of light which can be transmitted very efficiently and cheaply through optical fibres. Ultraviolet radiations can be passed in quartz rods in the same way. The shorter wavelengths of ultraviolet B and C radiations lead to even stronger refraction than occurs with light. The effect is used in directing ultraviolet radiations into an open wound or cavity with a quartz applicator mounted on a Kromayer lamp (see *Electrotherapy Explained*, Chapter 15).

If the block of glass illustrated in Figure 8.5 had been made with non-parallel sides, the emerging refracted beam can be made to pass in a different direction to the incident beam. Thus if the block were triangular, a prism, the beam can be made to refract twice in the same direction (Fig. 8.9a), giving a marked change to the direction of the beam and splitting a beam of white light into its constituent colours. This occurs because the shorter-wavelength violet rays, at around 400 nm, are more refracted than the longer-wavelength – around 750 nm – red rays. In fact it was the use of a prism to split the sunlight streaming through a hole in his window blind that led Isaac Newton in 1666 to experiment and discourse on the nature of light. It was Newton who used the word spectrum to describe the band of colours which appeared mysteriously out of white light like a ghost or spectre: hence the use of the same Latin root *spectare* to look at.

The ability to change the direction of light is exploited in all kinds of optical instruments. Figure 8.9b illustrates how a biconvex lens can alter the direction of a bundle of light rays. It is easy to see the principle on which it works if the lens is thought of as being made up of an infinite series of prisms. Clearly this biconvex lens causes a parallel bundle of rays to focus at a point, a converging lens, but all kinds of other configurations can be arranged. Thus all kinds of optical devices use simple or usually compound lenses made of different kinds of glass. These include the simple magnifying glass, microscopes, telescopes and

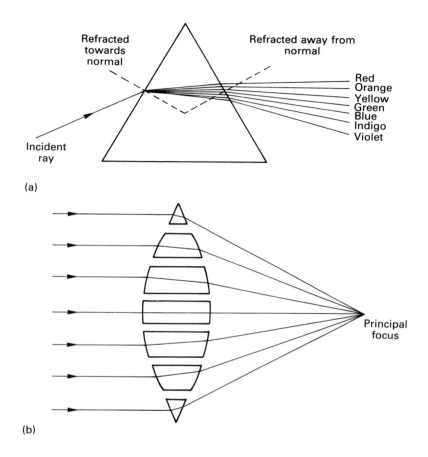

Fig. 8.9 (a) A prism and (b) a biconvex converging lens considered as a set of prisms.

binoculars, cameras and the lenses of spectacles and contact lenses. These are almost always (except contact lenses) glass lenses and thus have a shape fixed during manufacture. They therefore have a fixed focus and focusing of objects at different distances must be done by moving the lenses relative to one another. Thus microscopes and telescopes are focused by mechanical movement of one lens complex. In the human eye there are two parts involved in focusing light – the cornea and the lens. Instead of altering the distance between the lens and the retina, which is how a camera would focus an object on the photographic film, the lens is able to alter its shape. The cornea has a refractive index close to that of water, 1.33, but the lens is slightly higher, varying from about 1.386 at its edge to near 1.4 at its centre and is able to change shape by means of ciliary muscles in order to focus objects at different distances. This process is called *accommodation*.

Diffraction

Having considered the direction in which radiations travel when going from one medium to another it is also necessary to consider what effect an obstruction – a medium in which rays do not travel – has on radiations. In considering light it is not evident that the radiations bend around objects in any way because sharp-edged shadows are always cast. However, radiations of longer wavelengths, say, medium-wave radio, can clearly spread around hills or buildings. (The point was noted in connection with sound and ultrasonic waves in Chapter 6.) The effect of an obstruction or barrier depends on the wavelength of the radiation. Water waves passing through an opening in a barrier can be seen to spread out as they travel beyond it into the areas behind the barrier – the 'shadow' areas. This effect is called *diffraction* and the behaviour of waves beyond the gap in the barrier can be calculated from the way waves interfere and interact with one another. For the case of a gap in a barrier through which waves are passing the size of the diffraction effect can be shown to depend on the ratio of the width of the gap, d, to the wavelength, λ. If $\lambda : d$ is large a greater proportion of the wave energy spreads into the shadow of the barrier than when it is small. Thus for radio waves of several hundred metres, going through a 1 m doorway the effect is large, whereas visible radiations of about 500 nm (5×10^{-7} m) going through the same doorway show a negligible spread, hence the hard shadows. Visible radiations do, however, show diffraction provided the gap is made sufficiently small; in fact, all waves behave in this way. The effects can be demonstrated quite dramatically with a set of very narrow slits, a diffraction grating. Diffraction occurs whenever a wavefront is interrupted by an obstacle of any sort. The effect can often be seen when waves on the seashore are sweeping across some obstruction, like the end of a breakwater. If the obstructions are relatively small compared to the wavelength, such as tiny particles in the atmosphere, the effect is to produce circular wavefront disturbances and is called scattering. This effect causes the sky to look blue since the scattering is much more effective for the shorter wavelengths, hence more blue is scattered. In fact, it is the molecules of the atmosphere that give this effect. Larger particles, such as suspended dust, also cause scattering but of a different type. They act like tiny mirrors reflecting radiation in all directions.

Penetration and absorption

The relationship between the distance that electromagnetic radiations travel in a material and the amount of that radiation that is absorbed is of great importance and often not easy to specify. For radiation travelling in a homogeneous medium the amount of radiation absorbed at any point will be a fixed proportion of the total radiation at that point. Thus as the radiation progresses it becomes a smaller and smaller quantity so that the absolute amount – but not the relative amount – of radiation absorbed becomes less and less. In this way both the amount of radiation penetrating and the amount absorbed will fall exponentially with distance – distance being the depth of penetration. This exponential change is illustrated in Figure 3.4, showing that there is no point at which all the radiation is totally absorbed. To describe this pattern of absorption a single figure can be used which may be either the *half-value depth* or the *penetration depth*. As indicated in Figure 3.4, the depth or distance at which 50% of the radiation has been absorbed is called the half-value depth while the penetration depth refers to the point at which 63% of the radiations have been absorbed. Thus at successive half-value depths the radiation will fall to 50%, 25%, 12.5%, 6.25%, 3.12%, and so forth and at successive penetration depths the radiation will fall to approximately 37%, 13.7%, 5.07% and 1.87%. The value of the half-value or penetration depth can be expressed as this single figure for any given radiation in a particular material and can be very different for different wavelengths. Therefore, materials of a particular nature and suitable thickness can be used as filters. Coloured sunglasses are a simple example – many of the orange and yellow wavelengths are filtered out to leave relatively more green or blue radiations. Invisible ultraviolet radiations can also be filtered out.

A global example of this is the filtering of radiation from the sun by the earth's atmosphere. The sun, which is at a very high temperature, emits all kinds of radiation, some of which falls on the earth's atmosphere, to be reflected, absorbed or penetrate to the surface of the earth. Almost all of the short ultraviolet radiations below 300 nm are absorbed by ozone in the upper atmosphere with oxygen and nitrogen also absorbing some visible and ultraviolet radiations. Water vapour and carbon dioxide absorb much infrared radiation. All of these absorb certain wavelengths particularly strongly (Holwill and Silvester, 1973). The net effect of atmospheric filtration of solar radiation is to allow a band of radiations to penetrate to the earth's surface from about 300 nm in the ultraviolet band to 2000 nm in the infrared, with a peak around 650 nm. Destruction of atmospheric ozone due to the release of chlorofluorocarbons will allow the penetration of more short ultraviolet rays, and the continued release of carbon dioxide and other gases will increase absorption of reflected infrared radiation, contributing to global warming – the greenhouse effect.

Radiation applied to the tissues would follow the pattern outlined above only if the tissues were homogeneous, which they certainly are not. Thus there is reflection, refraction and scattering of radiations as they pass through the tissues, which has the general effect of causing more radiation to be absorbed close to the surface than would be predicted from a simple application of the penetration depth. Further, different tissues have very different penetration depths, making predictions even more uncertain. None the less, taking penetration depth to assess the site of tissues absorbing energy is useful.

THE EMISSION OF RADIATION FROM MATTER

So far, it has been sufficient to describe electromagnetic energy as the emission of photons of many different energies. It is necessary to consider how those photons are generated and absorbed to elucidate the further nature of radiation.

Energy levels and lined spectra

If an element in a gaseous form is given energy, say by heating it or by passing a current of electrons through it, the atoms can be made to emit light. This is in the form of a number of defined lines at specific frequencies characteristic of the particular atom. The fact that these radiations only are emitted indicates that the atom is capable of emitting or absorbing energy in particular amounts only. Thus there are particular 'allowed' states of an atom in which the electron orbits or orbitals can have more energy than in their most stable or ground state. When this happens the atoms are said to be excited to a higher energy level. The atom has absorbed a specific quantity of energy. Similarly, when it emits energy it does so in specific quantities, that is, a photon of a particular size is given off as the electron changes its orbital to a lower energy level (Figs 8.10 and 8.11 and see Fig. 2.3). (The term orbital is preferred to orbit because it implies the more subtle concept of the electron having an average pattern of motion around the nucleus but whose exact location cannot be specified.)

The energy changes are called *transitions* and correspond to a named series of lines in the hydrogen spectrum. The largest energy changes occur when electrons are moved between the ground state and other energy levels in the Lyman series of transitions. The photons (quanta) emitted are of high energy, i.e. higher frequency and shorter wavelength, and are, therefore, all in the ultraviolet region. They have energies between 10.2 and 13.6 eV (see Fig. 8.10 and Table 8.1). Lower energy transitions for hydrogen occur in the visible region, the Balmer series, and have energies between 1.9 and 3.4 eV. It will be seen from Table 8.1 that 1.9 eV is very close to the lowest energy of visible radiation which is at the red end of the spectrum. Figure 8.10 also shows that there are several other series of energy levels in which transitions can occur. The level labelled 0 eV in Figure 8.10 is the level at which the electron has just enough energy to leave the atom, i.e. the ionization energy. In the case of the hydrogen atom it takes more than 13.6 eV of added energy to remove its electron.

The hydrogen atom with a single electron is the simplest case but the same principle applies for all elements. The larger atoms are very much more complicated. The central point is that all atoms have a certain number of electron orbitals and hence a definite number of energy levels between which transitions can occur (see Chapter 2 and Fig. 2.3).

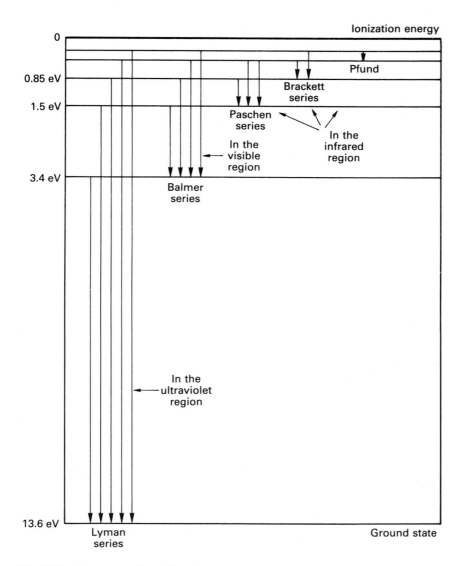

Fig. 8.10 An energy level chart.

If free atoms in a gas are made to collide with one another or with ions or electrons by means of an electric force, energy can be given to the atom raising an electron from the ground state to a higher energy level. This electron subsequently returns to a lower energy level (not necessarily the one from which it originated), releasing its energy as a photon of a frequency and wavelength specific to the transition energy. This is what occurs in a fluorescent lighting tube or an ultraviolet lamp. The electrons involved are those in the outer atomic orbitals. For example, mercury atoms give off specific 'lines' of radiation (lined spectra) in the

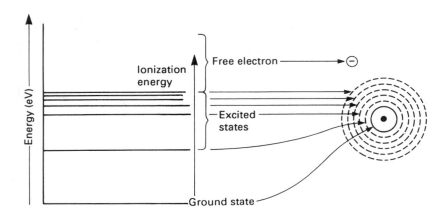

Fig. 8.11 An electron energy level diagram for hydrogen.

blue and ultraviolet regions. To achieve radiation of X-rays, transitions must be made to occur in the inner electron orbital of atoms to provide the much higher energy levels needed.

As described in Chapter 2, atoms combine with one another to form molecules, often sharing electrons so that they form molecular orbitals. These give rise to many more allowed energy levels so that there are more energy transitions possible and hence more radiations of different wavelengths can be emitted.

Molecules themselves are not static bodies but are able to move and alter their shape. Being an assembly of atoms, the molecule is able to vibrate as its constituent atoms move to and fro. This vibration can be of several kinds. In many molecules, the constituent atoms, a pair of carbon atoms, for example, simply move apart and come together again in a stretch type of vibration. With others, some atoms may swivel round other atoms – scissor deformation. The number of possible types of vibration depends on the number of atoms that make up the molecule; the more atoms, the more possible vibrations. Each of these possible vibrations has a specific frequency which depends on the masses of the atoms and the strength of the bonds joining them. These vibrations are also the motion of electrons and lead to the emission of electromagnetic radiations, notably in the near infrared.

Molecules are also able to rotate as a whole, the amount of energy involved depending on their size and mass. Again, vibration or oscillation of electrons is occurring which can lead to the emission of radiations. For many small, light atoms the radiation is in the far infrared bordering the microwave region.

In atomic gases, then, energy is emitted from a relatively few transitions, giving rise to a lined spectrum in the visible and ultraviolet regions. In molecular gases, the additional transitions produce a banded spectrum and extend to the infrared region. In solids and liquids the possible transitions are greater still and can have a much greater range of values, leading to a wide range of wavelengths close together and overlapping so that a continuous spectrum is produced.

THE PRODUCTION OF ELECTROMAGNETIC RADIATIONS

It has already been explained in principle that electromagnetic radiations are produced as a consequence of accelerating and decelerating electrons. How this occurs for the different radiations will now be considered.

Radio waves

Radio waves are produced by electrons oscillating to and fro in oscillating circuits such as those described in Chapter 4. Oscillations of low frequency, such as the mains alternating current at 50 Hz, produce radiations of negligible energy, but at higher frequencies, as shown in Table 8.1, radio waves are produced. With increasing frequency and shorter wavelength the radio waves behave more like a ray or beam. Thus longwave radio can easily diffract around mountains while shortwave television signals need line-of-sight transmission. Radio waves pass easily through the body tissues but at high intensities, such that high-frequency oscillating currents themselves are generated in the tissues, heating is produced. This occurs in shortwave diathermy (see Chapter 5). If average intensity is much reduced by pulsing the shortwave output then therapeutic benefits in increasing rates of healing occur (see *Electrotherapy Explained*, Chapter 10, pages 246–255). The major use of radio waves is, of course, for communication by picking up the transmitted radio frequency with a radio receiver, tuned to exactly the same frequency.

Microwaves

Microwaves are again produced by the oscillation of electrons in electrical circuits but since the frequency is higher the electrons must be made to oscillate very rapidly. This can be done with a device called a cavity magnetron. This is essentially a large thermionic valve (see Chapter 4) with a centrally heated cathode surrounded by a block of metal in which a number of very precisely sized cavities have been cut. A strong magnetic field is applied which causes electrons to travel spirally at high velocities in their passage from the cathode to the positively charged metal block. The high-velocity electrons passing across the cavities cause high-frequency oscillating currents in the cavities of a microwave frequency (see Table 8.1) and these are fed along a coaxial cable to an antenna or radiating aerial.

Microwave radiations can be picked up by suitably tuned circuits and are widely used for detecting the position of aircraft, ships, missiles, military targets and many other objects as radar. The short wavelength means that the radiation behaves more like a ray or beam and is easily reflected from metal surfaces. When used for communication, the transmitter and receiver stations have to be in line of sight. As already noted, microwave radiation is quite strongly absorbed by water but it is able to penetrate the tissues to some extent. A half-value depth of 3 cm is often quoted, but see *Electrotherapy Explained*, page 277.

Infrared radiation

All heated bodies emit infrared radiation as a continuous spectrum of many wavelengths (see Chapter 7). As explained, adding energy by heating leads to a vast number of transitions of a whole range of different energies. All objects are emitting and absorbing some infrared all the time. Whether they gain or lose heat energy will depend on the balance. As shown in Table 8.1, infrared ranges from about 100 to 0.77 μm. (N.B. There is no discontinuity between microwave and infrared; it is simply a matter of different definitions – long infrared and short microwave are the same.)

There is an important difference between infrared and microwave and radio waves. In the case of infrared the electron accelerations are due to random, natural processes so that the emitted radiations are random. There are photons of a whole range of energies, given out in separate trains, out of phase with one another and non-polarized to give a continuous spectrum. This is approximately illustrated by the output of an infrared lamp, illustrated in Figure 8.14 (see page 215). Note that the peak wavelength decreases with temperature and that the maximum emission wavelength is not the midpoint of the radiations but is displaced towards the shorter wavelength. (See *Electrotherapy Explained*, pages 288–290, for a description of therapeutic infrared lamps.)

Visible radiations

Visible radiations occur as a result of further heating, displacing the peak wavelength into the visible region of the spectrum. As the temperature increases, the wavelength of the peak radiation shortens. Thus at around 400°C all emitted radiation is infrared, at 700°C some red visible is being emitted as well, while at around 1500°C white light is emitted. This occurs because of transitions in outer orbital electrons.

They can be produced by electric discharges in fluorescent tubes, as described above, without significant heating. The therapeutic use of visible radiation is largely, but not entirely, confined to its laser form (see below).

Ultraviolet radiations

Ultraviolet radiations are again produced by the energy transitions of orbital electrons of rather more energy than those producing visible radiations (Fig. 8.10) – the Lyman series. In this case atoms and ions collide in a low-pressure gas causing an electron transition to occur which subsequently leads to the emission of a photon typical of that particular transition. (See *Electrotherapy Explained*, Chapter 15, for details of the production of therapeutic ultraviolet radiation and its effects on the human body.)

X-rays

X-rays result from energy transitions in the inner orbitals. These are produced by the very rapid deceleration of electrons which are given very high energy. A vacuum tube like a large cathode ray tube is used (see Chapter 4) and streams of electrons are directed from the cathode on to a tungsten anode mounted on a copper block. Very high energies are used – 100 kV or more – to accelerate the electron beam to produce sufficient X-rays for use. It is well known that this radiation is not only very penetrating but also causes tissue damage in sufficiently large doses.

Gamma rays

Gamma rays are given off from the nuclei of atoms which have emitted a particle in a reaction involved in radioactivity. The radiations are very energetic and even more penetrating than X-rays. They have similar, but more potent, destructive effects on tissues.

Cosmic rays

Cosmic radiation, which is not indicated in Table 8.1, consists largely of subatomic particles found in outer space, probably emanating from the sun. It is very energetic and interacts with the outer atmosphere so that it is not found on earth.

It will be noticed that there is a more or less abrupt shift in the changes that electromagnetic radiations are able to effect at the visible region. All the longer, low-energy, radio, microwave and infrared radiations are able to produce heating effects but only the very short – infrared, visible and beyond – are able to provoke chemical changes. Radiations with sufficient energy to disrupt the outermost electrons of atoms, turning the atom into an ion, such as gamma and X-rays, are called *ionizing radiation*.

Fluorescence

When a molecule of a solid or liquid is given energy so that an electron is raised to a higher level (Fig. 8.11), it can lose some energy in small quantities to neighbouring atomic molecules by many collisions, a process called radiationless transition because no photon is emitted. From this point the electron may fall to its ground state, emitting a photon in the usual way. If the original energy is given by a photon the emitted photon will have less energy. This is the process of fluorescence in which photons of short wavelength and higher energy, such as ultraviolet, are absorbed and a photon of longer wavelength and lower energy, such as visible radiation, is emitted. This is exactly what occurs in the coating of a fluorescent lighting tube.

Lasers

Another way in which the excited electron may lose energy is via a state in which the molecule or atom is very nearly stable so that the electron spends a relatively long time in what is called a *metastable state* before making the transition to the ground state. Electrons remain in higher-energy, excited states for a very short time, about 10^{-8} s, before falling to a lower level and emitting a photon, but they can remain in metastable states for much longer average periods, e.g. about 10^{-3} s. Electrons may leave this metastable state spontaneously or be triggered by a photon of precisely the same energy to fall to the ground state. When this happens it is called *stimulated emission*. In most circumstances there will be a high proportion of molecules in the ground state, in which case an incoming photon would cause the electron to gain energy up to the metastable level.

This allows a rather special type of radiation to be produced, called light amplification by stimulated emission of radiations, hence the acronym laser. This is special in that it is monochromatic, i.e. of one wavelength only. It is also coherent, that is, all the peaks and troughs of the magnetic and electric fields occur at the same time (temporal) and travel in the same direction (spatial). As a consequence of this spatial coherence and the way in which the laser radiation is produced the rays are in a narrow, parallel beam.

Non-coherent infrared and visible radiations are like a crowd of people all in different clothes, walking in different ways and out of step. Laser radiation is like a column of soldiers all marching in step (in phase), wearing the same uniform (monochromatic) and going in exactly the same direction (spatial coherence).

Lasers can be produced from solids, liquids and gases under the right conditions. One of the early lasers used a synthetic ruby rod, about 10 cm long and of 1 cm diameter with a xenon flash tube wound round it in a spiral (Fig. 8.12). The ends of the rod are flat and silvered to act as reflectors but one is made only partially reflecting. A powerful light flash from the xenon tube provides photons with energy to excite the molecules of the ruby rod to high energy levels. After a very short time in the high-energy band many spontaneous transitions occur to a metastable state in which the molecules remain for much longer periods. Thus there are more molecules in a metastable state than escaping from it to the ground level. This is the reverse of the normal, in which most molecules are in the ground state, so it is called a *population inversion*. When a transition occurs to the ground state a photon with a wavelength of 694.3 nm is emitted. This has exactly the right energy either to

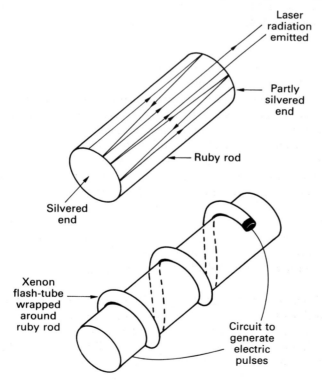

Fig. 8.12 A ruby laser.

raise a molecule to the metastable state or to stimulate one in the metastable state to fall to the ground state. Since there are many more of the latter, there is further stimulated emission of identical photons causing a cascade effect, i.e. one photon releasing another identical one then these two stimulating two more and so on (Fig. 8.13). The radiations, being reflected up and down the rod, quickly affect all the molecules to bring them to their ground state, thus a pencil of red light escapes through the partially reflecting end. This all occurs in a very brief time so that the short pulse of coherent, monochromatic, red light (a pulsed laser) can have very high power.

Other materials can be made to operate in a similar way. Helium-neon lasers consist of a tube containing these gases at low pressure and surrounded by a flashgun tube. Excitation leads to different energy levels between these two atoms and a transfer of energy giving off a photon of a wavelength equal to the energy gap of 632.8 nm (Fig. 8.14). Carbon dioxide is used to provide high-intensity (up to 20 000 W) infrared lasers for, among other uses, the surgical destruction of tumours. Chemical lasers, in which the energy source is a chemical reaction, have also been developed and are notably more efficient than ordinary lasers which only convert around 2% of their energy to laser radiation. Organic lasers, in which a complex organic dye is utilized, can be made and tuned so that they produce radiations at a predetermined wavelength.

Semiconductors are extensively used for this purpose. There are various kinds involving gallium aluminium arsenide (GaAlAs). As semiconductors (see Chapter 4) these materials can be made as diodes. The stimulation is effected by an electric potential affecting electron–hole pairing in the crystal lattice. This effectively stores energy in the material which can be released in the form of identical photons in the manner described. The photons are reflected to and fro in the material, to be ultimately emitted as a laser beam. These devices are relatively cheap to construct, quite robust, can be made to be very small and emit a

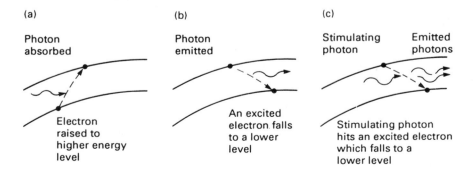

Fig. 8.13 Absorption and emission of photons: (a) absorption; (b) emission; (c) stimulated emission.

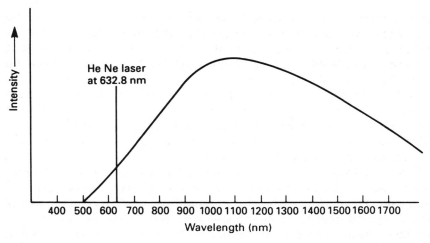

Fig. 8.14 Comparison of an infrared laser output with an infrared lamp.

predetermined wavelength by varying the relative proportions of gallium to aluminium. Some examples are given in Table 8.6.

The laser mechanism described so far would lead to single bursts of radiation and hence is referred to as pulsed laser. It allows very large amounts of energy to be released in a very short time, in the order of 10 kW mm^{-2} (Holwill and Silvester, 1973). The relatively long intervals between pulses ensures that the average power is often quite trivial. Continuous lasers can be constructed by giving the electrons of the laser material energy, as described, which pumps them to a higher energy level. The stimulating photons are, however, of a lower energy and the electrons are able to fall to their ground state through two transitions.

Table 8.6 Examples of lasers

Laser type			Wavelength (nm)	Radiation
Ruby			694.3	Red light
Helium-neon			632.8	Red light
Gallium aluminium arsenide diodes	Continuous wave		650	Red light
			750	Red light
			780	Infrared
			810	Infrared
			820	Infrared
			850	Infrared
			1300	Infrared
	Pulsed injection		860	Infrared
			904	Infrared
Carbon dioxide			10 000	Infrared

Thus pumping and stimulated emission can occur at the same time and as the processes will not interfere with one another a continuous output results. It is, of course, possible to pulse the output of a continuous laser electronically, giving pulses of the required length and a variable frequency.

Measurement of laser output

Lasers used in physiotherapy are variously called low-level, mid-power or soft; see Table 8.7, which also indicates some of the more common uses of lasers. Most lasers used in physiotherapy have an output power in the tens of milliwatts. The total output power is often specified but, as this is spread over the whole beam area, the energy per unit area is given by watts per square centimetre, sometimes called the power density. The total energy emitted will depend on the time for which the power is applied (1 W being 1 J s^{-1}), so that a 10 mW laser applied for 100 s would give 1 J of energy. If this energy were delivered with a beam of 10 mm^2 cross-sectional area, it would give an energy density of 10 J cm^{-2}. As with other radiations and ultrasound, this expresses the energy applied, not what is absorbed which is the true dosage. If the laser is pulsed, the temporal peak power must be distinguished from the temporal average power which depends on pulse length, pulse frequency and pulse power. King (1989) points out that the pulse is not necessarily square-wave, so the peak power is only attained for a fraction of the total pulse length, making average power difficult to calculate.

The effects of lasers on the tissues and their therapeutic uses are described in *Electrotherapy Explained*, pages 307–310. The mechanism and extent of its beneficial effects are not yet firmly established.

Table 8.7 Classification of lasers

Effect	Number	Range of power	Usage
No effect on eye or skin	1 2	Low power	Blackboard pointer Supermarket barcode reader
Safe on skin, not on eye	3A 3B	Mid-power	Therapeutic–physiotherapy models; up to 50 mW mean power (physiotherapy with 3A and 3B lasers is also called low-level laser therapy; LLLT)
Unsafe on eye and skin	4	High power	Surgical–destructive

Appendix A:
Power of 10 notation

This is a method of making the handling of large numbers much simpler. Basically it obviates the need to read and write long strings of noughts and allows multiplication and division to be done by addition and subtraction, which is much easier. It is also known as scientific notation. The basis is set out in Table A.1. The small superscript number to the right of the 10 is called the exponent.

Table A.1 Power of ten notation

10^{-4}	0.0001	one-ten-thousandth
10^{-3}	0.001	one-thousandth
10^{-2}	0.01	one-hundredth
10^{-1}	0.1	one-tenth
10^{0}	1	one
10^{1}	10	ten
10^{2}	100	a hundred
10^{3}	1000	a thousand
10^{4}	10 000	ten thousand
10^{5}	100 000	a hundred thousand
10^{6}	1 000 000	a million
10^{7}	10 000 000	ten million
10^{8}	100 000 000	a hundred million
10^{9}	1 000 000 000	a billion*

*Almost universally, the word billion refers to 1 000 000 000 (one thousand million) or 10^9. Formerly in the UK the word meant 1 000 000 000 000 (one million million) or 10^{12}. In the USA and increasingly in the UK, this is referred to as a trillion.

Notice that in Table A.1 the numbers have sensible meanings. Thus 10 squared (10^2) is 100 and similarly the negative powers are *reciprocals* of the numbers, so that $10^{-2} = 1/100 = 0.01$. Notice also that the power of 10 notation is much shorter to write than the equivalent numerical or word representation for all but the shortest positive numbers. The larger the number, the more convenient the system becomes – Table A.1 could be extended both upwards (negative) and downwards (positive).

The value of any given power of 10 notation can be found in another way by considering the exponent to indicate the location of the decimal point. After putting down the initial 1, the place of the decimal point is determined by the number of noughts given by the exponent, to the right for positive and the left for negative (Table A.2).

So far the power of 10 notation has been considered only with regard to 10, 100, 1000 and so forth, but any number can be described by the system. When 10^2 is written it is the same as 1×10^2 and means 100.

Physical Principles Explained

Table A.2 Power of ten notation

−4	−3	−2	−1	0	1	2	3	4
↓	↓	↓	↓	↓	↓	↓	↓	↓
0	0	0	0	1	0	0	0	0

So:

10^{-4}	move decimal point 4 places to left of 1 =	0.0001
10^{-3}	move decimal point 3 places to left of 1 =	0.001
10^{-2}	move decimal point 2 places to left of 1 =	0.01
10^{-1}	move decimal point 1 place to left of 1 =	0.1
10^{0}	move decimal point 0 places from 1 =	1
10^{1}	move decimal point 1 place to right of 1 =	10
10^{2}	move decimal point 2 places to right of 1 =	100
10^{3}	move decimal point 3 places to right of 1 =	1000
10^{4}	move decimal point 4 places to right of 1 =	10 000

Similarly, 2×10^2 would be 200 and 2.4×10^2 would be 240. In this way any number can be written as the digit or digits times some power of 10. Some examples from the text are given in Table A.3.

What has been described up to this point is a convenient method of describing numbers but the real value and strength of the notation lie in the ability to simplify calculations. In order to multiply two numbers together, all that is necessary is to add the exponents, for example:

$$10^2 \times 10^3 \ (100 \times 1000) = 10^5 \ (100\,000) \text{ because } + 2 + 3 = 5$$

$$10^2 \times 10^{-4} \ (100 \times 0.0001) = 10^{-2} \ (0.01) \text{ because } + 2 - 4 = -2$$

$$10^2 \times 10^{-2} \ (100 \times 0.01) = 10^0 \ (1) \text{ because } + 2 - 2 = 0$$

This explains why 10 can be taken as 1.

In the same way, division of numbers can be effected by subtracting the exponents, for example:

$$\frac{10^4}{10^2} = \frac{10\,000}{100} = 10^2 \ (100)$$
$$(\text{because } + 4 - 2 = + 2)$$

or

$$\frac{10^{-1}}{10^2} = \frac{0.1}{100} = 10^{-3} \ (0.001)$$
$$(\text{because } - 1 - 2 = -3)$$

Table A.3 Examples of power of ten notation from text

Approximate velocity of radiation *in vacuo*	$= 3 \times 10^8$ m s^{-1} (300 000 000 m s^{-1})
Approximate velocity of sonic waves in tissues	$= 1.5 \times 10^3$ m s^{-1} (1500 m s^{-1})
Energy of electron volt	$= 1.6 \times 10^{-9}$ J (0.0000000016 J)
Energy of 1 Calorie (or 1 kilocalorie)	$= 4.18 \times 10^3$ J (4180 J)

Only the exponents can be added or subtracted; the digits are multiplied or divided as usual. For example, to calculate how long it takes an ultrasound beam travelling at 1.5×10^3 m s^{-1} to pass 4.5×10^{-2} m through the tissues:

$$\frac{4.5 \times 10^{-2}}{1.5 \times 10^{-3}} = 3 \times 10^{-5}, \text{ i.e. } 0.00003 \text{ seconds}$$

Another example: if we wish to calculate the total energy, in joules, emitted by a 250 W infrared lamp applied for a 20 minute treatment, as considered in Chapter 1:

250 W × 1200 seconds gives (2.5×10^2) W × (1.2×10^3) s

So $2.5 \times 1.2 = 3$ and $10^2 + 10^3 = 10^5$ and gives 3×10^5 or 300 000 J.

Appendix B: A fairytale

Once upon a time a miller lived with his daughter on the slopes of a hill near a fast-flowing river. Uncertain winds made it difficult for the miller, who wished he had a more reliable way to turn the sails of his windmill. One day while brooding over his troubles besides the river, where it splashed its way over a pretty little waterfall, he was struck by the idea of using the fast-flowing water to turn the mill. He therefore fitted a waterwheel to the side of his windmill and dug a channel from the riverbank under the waterwheel and back to the same place on the river bank, as pictured in Figure B.1. The channel filled but no flow of water occurred so that the waterwheel would not turn and the miller was gravely disappointed. The still water in the channel, however, attracted a number of frogs and the miller's daughter took to kissing them in the belief that one might turn into a handsome and wealthy prince whom she could marry. None did: in fact, the frogs disliked the whole activity, especially the slower ones who were repeatedly kissed because the miller's daughter was quite unable to distinguish those frogs that had already had the treatment from those that had not.

One day a vagabond was given overnight lodging by the miller. Hearing about the miller's troubles during the evening, he decided to

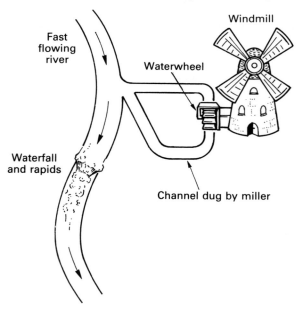

Fig. B.1 The miller's channel.

repay the hospitality. The next morning the vagabond got up early, took a spade, filled in the old channel and dug a new one to a point on the river bank below the waterfall, as shown in Figure B.2. Immediately a strong current of water flowed through the channel, turning the waterwheel and grinding the corn.

The miller was delighted and rapidly became very wealthy. The miller's daughter abandoned osculation with the frogs and married the vagabond who took up the study of hydrological engineering. They all lived happily ever after.

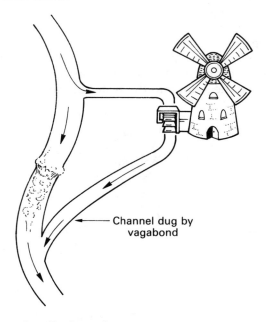

Channel dug by vagabond

Fig. B.2 The vagabond's channel.

Appendix C:
A water oscillator

In order to understand more clearly the way in which an electrical oscillating system works, an analogy can be found with a hydromechanical system. Such a system has no practical use and has probably never been built but it is easy to imagine how it would work and thus see how the electrical oscillating system operates.

The arrangement is shown in Figure C.1. The large container is divided into two compartments by an elastic rubber membrane: the two compartments are connected to each other by a pipe. The paddles of a large heavy waterwheel dip into the water in the pipe.

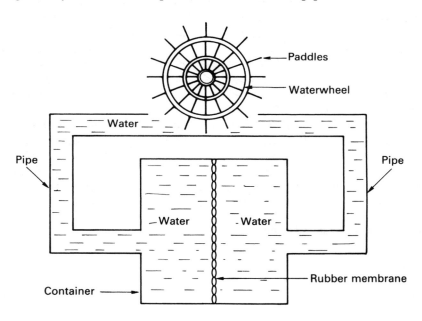

Fig. C.1 A hydromechanical oscillating system.

Now suppose the waterwheel is turned by hand in a clockwise direction: it will sweep water in the pipe – pump the water – from the right to the left side of the container. This will overfill the left-hand compartment, stretching the elastic dividing wall so that it bulges to the right. If the wheel is now released, the elastic membrane will recoil, pushing water from the left-hand compartment along the pipe into the right-hand one. The flow of water in the pipe will turn the waterwheel anticlockwise, slowly at first because it is heavy. The waterwheel will continue to turn because of its momentum, even after the rubber membrane has recovered to its central position and thus ceased to push the water. The continuing anticlockwise motion of the waterwheel will now pump water from the left- to the right-hand compartment, now overfilling the right side and stretching the membrane to the left.

The process will now repeat the cycle over again and continue to do so but at each oscillation some energy is lost as friction in the wheel and water movement. Thus each cycle will have a little less amplitude than the previous one – that is, less water will move – but the time taken by each cycle will be identical. The system will exhibit damped oscillations, as shown in Figure 5.3.

The large container is the equivalent of the capacitor in an electrical oscillating circuit with the rubber membrane as the dielectric. The water – equivalent to electrons – is unable to pass through the rubber barrier but is able to distort it and energy is stored by this distortion, just like the dielectric of the capacitor. Recoil of the membrane pushes the water through the pipes, like electrons being driven through the circuit. The movement of the waterwheel is comparable to the magnetic field, increasing slowly at first because of its inertia, equivalent to the back e.m.f. The continued rotation of the wheel due to momentum is the forward e.m.f. which recharges the capacitor in the other direction in the same way as it forces water out of one and into the other compartment.

The friction losses of the mechanical system are equivalent to the losses due to ohmic resistance and impedance in the electric circuit.

If continuous undamped oscillations were required it would only be necessary to give the waterwheel a little extra turn or two at each oscillation to add the lost energy, but it would be necessary to do it at the right moment. This is just what the transistor or valve does in the circuits shown in Figure 5.4.

Appendix D:
To help the visualization of longitudinal waves

Sonic waves, as explained, are longitudinal waves consisting of an oscillatory motion of air molecules. The resulting compressions and rarefactions may be visualized with the aid of a simple model.

Crova's disc

A large circular piece of cardboard, about 25 cm in diameter, with a small circle about 1.5 cm in diameter drawn in its centre, is needed. Eight equidistant points are marked and numbered on the circumference of the small inner circle. Using these points as centres and starting at number one, a series of progressively larger circles are drawn on the card. Another sheet of card is placed over the large circle and pivoted to it at its centre point. A wide slit is cut in the covering card sited over the radius of the large circle. Now, if the circular card is turned continuously and the lines viewed through the slot, they will appear to come together and move apart, representing compression and rarefaction as they travel along the length of the slot. This looks very similar to the way layers of air molecules would move in a sound wave, if they could be seen. Compare this with Figure 6.1. As can be seen from the card, the lines are simply moving to and fro about a mean position, like the air molecules.

Appendix E: Table of small units

Name	Symbol	Relation to metre	Fraction of a metre
Metre	m	1	
Centimetre	cm	10^{-2}	one-hundredth
Millimetre	mm	10^{-3}	one-thousandth
Micrometre, micron	μm	10^{-6}	one-millionth
Nanometre	nm	10^{-9}	one-thousand-millionth
Ångström unit (not an SI unit)	Å	10^{-10}	one-ten-thousand-millionth
Picometre	pm	10^{-12}	one-million-millionth

Appendix F: Refraction

If it is not immediately obvious why a change of velocity when a wave enters a different medium causes it to bend, the following simple and slightly macabre analogy may be helpful.

Consider a group of pallbearers carrying a coffin across a field who come to the edge of a shallow muddy stream which they have to cross. Their line of slow march approaches the stream at an angle. The bearers on the side reaching the edge of the stream first will be slowed as they struggle through the mud. Those on the other side, still on dry land, can move faster so that the direction of the cortège changes. Once all are wading in mud the velocity of both sides is the same, so that the coffin again travels in a straight line. At the far edge of the stream the first side to emerge on to dry land immediately speeds up, so swinging the line of travel of the coffin back in the original direction. Figure 8.5 shows what happens to light rays passing from air through a parallel-sided block of glass but is analogous to the parallel-sided muddy stream. In both cases the velocity is diminished in the denser medium, causing the front to change direction. It can be imagined that if the muddy stream had had a triangular configuration, the direction of the cortège could be further altered in the same direction, like light radiations passing through a prism.

References

References in this book to *Electrotheraphy Explained* are to Low J. and Reed A. (1990) *Electrotherapy Explained: Principles and practice*. Butterworth-Heinemann, Oxford.

CHAPTER 1

Ritchie-Calder, Lord (1970) Conversion to the metric system. *Scientific American*, **223**.

CHAPTER 3

Ward A.R. (1986) *Electricity Fields and Waves in Therapy*. Science Press, Marrickville, NSW.

CHAPTER 4

Asimov I. (1966) *Light, Magnetism and Electricity*. Signet Science Library Inc, New York.
Shiers G. (1971) The induction coil. *Scientific American*, **224**, 80–87.

CHAPTER 5

Patterson R.P. (1983) Instrumentation for electrotherapy. In: Stillwell K. (ed.) *Therapeutic Electricity and Ultraviolet Radiation*, 3rd edn. Williams & Wilkins, Baltimore, MD, 65–108.

CHAPTER 6

Cromer A.H. (1981) *Physics for the Life Sciences*, 2nd edn. McGraw-Hill, New York.
Docker M.F. (1987) A review of instrumentation available for therapeutic ultrasound. *Physiotherapy*, **73**, 154–155.
Freeman I. M. (1968) *Sound and Ultrasonics*. Random House Science Library, New York.
Frizzell L.A. and Dunn F. (1982) Biophysics of ultrasound. In: Lehmann J. F. (ed.) *Therapeutic Heat and Cold*, 3rd edn. Williams & Wilkins, Baltimore, MD, 353–385.
Hekkenberg R.T., Oosterbaan W.A. and van Beekum W.T. (1986) Evaluation of ultrasound therapy devices. *Physiotherapy*, **73**, 390–394.

Lloyd J.J. and Evans J.A. (1988) A calibration survey of physiotherapy ultrasound equipment in North Wales. *Physiotherapy*, **73**, 154–155.

Ward A.R. (1986) *Electricity Fields and Waves in Therapy*. Science Press, Marrickville, NSW.

Williams R. (1987) Production and transmission of ultrasound. *Physiotherapy*, **73**, 113–116.

CHAPTER 7

Buetner K. (1951) Effects of extreme heat and cold on human skin. *Journal of Applied Physiology*, **3**, 703–713.

Cetas C.T. (1982) Thermometry. In: Lehmann J.F. (ed.) *Therapeutic Heat and Cold*. Williams & Wilkins, Baltimore, MD, 35–69.

Cromer A.H. (1981) *Physics for the Life Sciences*, 2nd edn. McGraw-Hill, Auckland.

Davison J.D., Ewing K.L., Fergason J. *et al.* (1972) Detection of breast cancer by liquid crystal thermography. *Cancer*, **29**, 1123.

Fricke J. (1993) The unbeatable lightness of aerogels. *New Scientist*, **1858**, 31–34.

Hardy, J.D. (1982) Temperature regulation, exposure to heat and cold and effects of hypothermia. In: Lehmann J.F. (ed.) *Therapeutic Heat and Cold*. Williams & Wilkins, Baltimore, MD, 172–198.

Nightingale A. (1959) *Physics and Electronics in Physical Medicine*. G. Bell, London.

Sekins K.M. and Emery A.F. (1982) Thermal science for physical medicine. In: Lehmann J.F. (ed.) *Therapeutic Heat and Cold*. Williams & Wilkins, Baltimore, MD, 70–132.

Togawa T. (1985) Body temperature measurement. *Clinical Physics and Physiological Measurement*, **6**, 83–108.

CHAPTER 8

Hay G.A. and Hughes D. (1972) *First Year Physics for Radiographers*. Baillière Tindall, London.

Holwill M.E. and Silvester N.R. (1973) *Introduction to Biological Physics*. John Wiley, London.

King P.R. (1989) Low level laser therapy: a review. *Lasers in Medical Science*, **4**, 141–150.

von Frisch K. (1967) *The Dance Language and Orientation of Bees*. The Belknap Press of Harvard University Press, Cambridge, MA.

Bibliography

Abbott A.F. (1984) *Physics*, 4th edn. Heinemann Educational Books, London.

Akrill T.B., Bennet G.A.G. and Millar C.J. (1979) *Physics*. Hodder and Stoughton, London.

Asimov I. (1969) *Understanding Physics*, vol. 3. *The Electron, Proton and Neutron*. Signet Books. The New American Library, New York.

Asimov I. (1987) *New Guide to Science*. Penguin Books, Harmondsworth.

Cromer A.H. (1981) *Physics for the Life Sciences*, 2nd edn. McGraw-Hill, Auckland.

Duncan T. (1977) *Physics for Today and Tomorrow*. John Wiley, London.

Faller J.E. and Wampler E.J. (1970) The lunar laser reflector. *Scientific American*, **222**, 38–49.

Griffin J.E. and Karselis J.C. (1988) *Physical Agents for Physical Therapists*, 3rd edn. Charles C. Thomas, Springfield, IL.

Hay G.A. and Hughes D. (1972) *First Year Physics for Radiographers*. Ballière Tindall, London.

Holwill M.E. and Silvester N.R. (1973) *Introduction to Biological Physics*. John Wiley, London.

Jolly W.P. (1972) *Electronics*. (Teach Yourself Books.) The English Universities Press, London.

Moritz A.R. and Henriques F.C. (1947) Studies in thermal injury II. The relative importance of time and surface temperature in causation of cutaneous burns. *American Journal of Pathology*, **23**, 695.

Richardson I.W. and Neergaard E.B. (1972) *Physics for Biology and Medicine*. John Wiley, London.

Swartz C.E. and Goldfarb T.D. (1974) *A Search for Order in the Universe*. W.H. Freeman, San Francisco, CA.

ter Haar G. (1978) Basic physics of therapeutic ultrasound. *Physiotherapy*, **64**, 100–103.

ter Haar G. (1987) Basic physics of therapeutic ultrasound. *Physiotherapy*, **73**, 110–113.

Uvarov E.B. and Isaacs A. (1986) *Penguin Dictionary of Science*. Penguin Books, Harmondsworth.

Ward R.R. (1986) *Electricity, Fields and Waves in Therapy*. Science Press, Marrickville, NSW.

White D.C.S. (1974) *Biological Physics*. Chapman & Hall, London.

Index

Absorption:
  electromagnetic waves, 34
  of waves, 31
Acoustic impedance, 146, 147, 195
Albutt, T. C., 164
Alternators, 77
  stators and rotors, 78, 80
Ammeters, 48, 69
Amplifiers:
  gain of, 125
  transistors as, 125
Amplitude of waves, 26
Anodes, 100
Antinodes, 31
Atomic masses, 12
Atomic nucleus, 10, 11
  hadrons in, 178
Atomic number, 12
Atoms, 9, 10–13
  amphitheatre model, 14
  bonding, 15, 21
  covalent bonding, 15, 21
  emitting light, 206
  forces of attraction, 19
  ground state, 14
  ionic bonding, 15, 21
  particles of, 11, 13
  planetary model of, 11
  polarization, 53
  quantum numbers, 13
  uncertainty model, 12

Baryons, 178
Batteries, for therapeutic apparatus, 115
Black, Joseph, 154, 156
Blood, thermal conductivity, 161
Body:
  core temperature, 169
  local cooling, 175
  therapeutic heating of, 175
Body isotherms, 170
Body temperature, 168, 169–76
  energy altering, 153, 175
  gain and loss, 171
  heat losss mechanism, 172, 174
  isotherms, 170
  maintenance of homeothermy, 171
  measurement of, 168
  outside energy affecting, 175
  physiological control of, 153, 173
  rectal measurement, 169
  zeroheat flow measurement, 169
Boyle's law, 18
Brown, Robert, 17

Brownian motion, 17, 18, 20
Buzzers and bells, 67, 68

Calculators, 168
Callan, N. J., 86
Calorie and kilocalorie, 8
Calories, 152
Capacitance, 51
Capacitance resistance circuits, 55, 116, 117, 118, 121, 122, 123
Capacitors, 51, 88
  impedance, 88
  resistance and, 59, 60, 61
  variable, 52
Carbon, 12, 10
Carbon atom, 10, 15
Carlisle, Sir Anthony, 99
Cathode rays, 87, 111
Cathode ray oscilloscope, 111
Cathodes, 100
Cavity magnetron, 209
Cells, see Electric cells
Celsius, Andreas, 149
Celsius scale, 149, 150
Centigrade scale, 149
Charles's law, 18
Chemical energy, conversion to electrical, 94
Chemical reactions, 15
Chips, 109
Choke coil, 89
Colour, 194–5
Compounds, 10, 15
Conductance, 47
Conduction, 101, 159
Conductors, 36, 37
  electron flow in, 36, 37
  self-induction, 75
  see also Semiconductors
Conservation of energy, 7
Conservation of mass, Law of, 158
Convection, 162
  forced, 163
  thermal, 163
Cooling, 162
Cosine law, 198
Cosmic rays, 33, 211
Coulomb, 39
Coulomb, Charles Augustine de, 179
Coulomb's law, 179, 182
Crova's disc, 234
Cryotherapy, 157
Crystal lattices, 21
  atoms binding in, 102
  electrons moving through, 40

· 231